INSTRUCTOR'S MANUAL

FOUNDATIONS OF CHEMISTRY IN THE LABORATORY

ELEVENTH EDITION

Morris Hein
Mount San Antonio College

Judith N. Peisen
Hagerstown Community College

Leo R. Best
Mount San Antonio College

Robert L. Miner
Mount San Antonio College

WILEY

JOHN WILEY & SONS, INC.

Cover Photo: Tim Davis/Corbis Images

To order books or for customer service call 1-800-CALL-WILEY (225-5945).

ISBN 0-471-46865-7

10 9 8 7 6 5 4 3 2

CONTENTS

NOTE: For your ease of use, the KEYS in Section E match the page numbering in the student lab manual (even though this does not match the actual page numbers of this manual). Therefore, the KEY for the Report for Experiment 1 (Section E) is on pages 9 and 10, and the KEY for the Report for Experiment 2, which follows the Report for Experiment 1 directly, starts on page 19, etc.

A. LABORATORY MANAGEMENT

This section is not intended to tell you how to manage your laboratory program. However, we have learned a lot over many years of working with beginning chemistry students and some practices work better than others. Some of these practices are included here.

1. At the beginning of each term,
   a. We give each student a copy of the schedule of experiments. Wherever possible we prefer to schedule experiments which support the ongoing lecture content. Copies of this schedule are also posted on the laboratory bulletin boards.
   b. A system for replenishing supplies as they run out should be in place to avoid last-minute problems with availability.

2. All sections of a given course do the same experiment during the same week insofar as possible. During weeks with holidays, this is accomplished by omitting an experiment for some sections or assigning a "Review Session" or "Recitation" for others.

3. We begin each session with a prelab discussion period. During this time we provide an overview of the experimental objectives, discuss principles, demonstrate techniques, emphasize safety and waste disposal precautions, and answer student questions pertinent to the current experiment. The length of these discussion periods averages about 25 minutes.

4. In order to make laboratory work more than an exercise in manipulation, we grade all report forms and schedule at least two lab exams per semester, a midterm and a final exam.

5. Report forms are collected for every experiment on the same day the experiment is performed. This encourages students to prepare before coming to each session, to budget their working time more effectively, and to focus on the underlying principles of the experiment as they answer questions and do the problems on the report form. Exceptions to this policy are made at the discretion of the instructor when the discussion, experimental work, or examinations consume more than the allotted time.

6. All report forms are graded using student readers when available. The answer keys (Section E) are setup to match with the student report forms and facilitate rapid evaluation. With this edition, we have provided sample data for all experiments. We require students to maintain a file of and use their graded report forms to study. Students are required to hand in these files at the end of the term (or when dropping the class).

7. We emphasize the proper disposal of wastes constantly and consistently right from the start. The procedures provide specific instructions every time a student is faced with a disposal decision.
    These instructions are emphasized by the waste icon (left margin). The waste collection containers we use for each experiment are listed in this Instructor's Manual in Section C (Notes on Individual Experiments at the end of each discussion) and in the lab manual, Appendix 8. Waste disposal is an important component of the Chemical Management Programs for our institutions, and is a line item in the department budgets. In the final analysis, each chemistry department will and should have their own regulations for disposal of wastes.

B. EVALUATION OF EXPERIMENTS

1. We try to grade and return report forms the week after each experiment is done. Each student response on the report form is assigned a point value: qualitative observations, properly recorded quantitative data, calculation setups and problem solutions, are all included. Unless this is done, some students begin to neglect the descriptive parts of the

experiment, get careless with significant figures in measurements, and neglect the calculation setups. When both descriptive and quantitative parts are included on a report form, the quantitative parts are generally weighted more heavily. Lab partners are encouraged to work together but to put all answers into their own words.

2. Answer Keys for all the experiments are given in Section E. We have included *sample* data for many experiments, thus student data will not match the key exactly. When student graders are used, we find they need instruction and/or supervision occasionally concerning their judgment of student answers.

3. It is not our intent to assign point values to parts of experiments. This evaluation is the prerogative of each instructor and will vary due to individual preferences and emphasis. Where our experience has contributed to concrete evaluation, we have given a grading scale and suggested limits for certain data. These suggested evaluations are included in Section C, Notes on Individual Experiments.

C. NOTES ON INDIVIDUAL EXPERIMENTS AND WASTE REQUIREMENTS

FIRST LABORATORY CLASS MEETING

At the first laboratory session we assign lockers and supervise the checking out of equipment by students. We discuss every item with its rationale in the Laboratory Rules and Safety Procedures in the lab manual emphasizing that no student can participate without eye protection. We point out special equipment and features of the laboratory and demonstrate special items of safety equipment. We go over waste disposal procedures and familiarize students with the location of waste containers. For the protection of the instructor, the school, and the student, it is becoming common practice for each student to sign a safety contract stating their knowledge and understanding of these rules. Thus, when this session is complete, every student signs our contract which is filed in the department office. A sample of this contract is provided below. If you choose to require a safety contract for your students, this sample can be copied or modified to suit your needs. If time permits during this first class meeting, Part of Experiment 1 is begun.

I have read and understand the Laboratory Rules and Satefy Procedures in my laboratory manual. I accept the responsibility for following all of these rules and procedures in order to maintain a safe laboratory environment for myself and others.

Course _____ Instructor _____

Student Signature _____ Date _____

Do you wear contact lenses? yes _____ no _____

EXPERIMENT 1: Laboratory Techniques

Since most of the students have had little or no prior laboratory experience, it is desirable to demonstrate the following techniques:

1. Adjusting the burner.
2. Cutting, fire-polishing, and bending glass tubing (specimens of properly fire-polished and bent tubing are made available for student inspection).

3. Inserting glass tubing into and removal from rubber stoppers. Care is taken to point out the safety advantages of (a) fire polishing, (b) proper use of a suitable lubricant (glycerol), and (c) the technique of gripping the tubing close to the stopper. NOTE: A No. 3 cork borer is a useful tool for removing 6 mm glass tubing from stoppers.

4. Folding filter paper to form a cone and seating the cone in a funnel.

5. Using a stirring rod in transferring a liquid (decanting from a beaker to a filter).

6. We like to have students see the actual MSDS sheets for the chemicals used in this experiment so they realize that the waste regulations have a valid rationale. This is not a formal part of the experiment, but the sheets are made available.

Note that the report calls for the instructor to evaluate student-prepared glassware.

Waste requirements:
Two waste containers are needed: (1) a bottle for Waste Heavy Metal, Pb^{2+}, and (2) a jar for filter paper with PbI_2.

EXPERIMENT 2: Measurements

We consider this a vital experiment for getting students off to a good start in several areas:

1. basic measurements and their relationship to significant figures
2. problem setups and dimensional analysis
3. the use of rules for significant figures in calculations
4. the difference between precision and accuracy in measurements, and
5. following directions

Each instructor will probably want to emphasize certain of these areas in the prelab discussion and report grading. Review of this experiment when the report is returned is particularly valuable. We refer students to Study Aids 1 and 5, and assign Exercises 1 and 2 in connection with this experiment. For help with the use of a scientific calculator, we refer them to Study Aid 4.

Since various types of balances are used in different schools, we have omitted detailed instructions on the use of the balance. The instructor should demonstrate weighing procedures. Any tips that will speed up the students' weighings will be most valuable in this and in many subsequent experiments. For economy of weighing time it is especially important to point out the difference between weighing an exact quantity (utmost precision) and weighing an approximate quantity.

The directions for the Eleventh Edition are written for an electronic balance with the uncertain digit in the 0.001 place. We also introduce the "tare" function and we require that all measurements be recorded to include one uncertain digit. Rounding is done only when a measurement is used in a calculation and the rules of significant figures are applied. The degree of precision in the answer keys must be adjusted for the equipment used by your students.

When the number of balances is limited, students are apt to be waiting in line to use the balance for Part B. We suggest that some students begin with Part B and others delay Part B until the balances are more available.

Unknowns:
We use metal slugs, 7/16 to 1/2 inch diameter by approximately $1\frac{3}{4}$ inches long. Slugs should be somewhat variable in length and should be numbered.

Materials for slugs: Aluminum, brass, and steel (other materials can be used).

Evaluation:
For this experiment, in particular, evaluation is made on (1) precision of data, (2) correct problem setups, (3) density determinations, and (4) units in Part E. No precision is checked for Parts C and D.

Permissible density ranges (suggested).

1. Water   0.94 – 1.06 g/mL
2. Rubber stopper   1.1   – 1.5   g/mL
3. Unknowns
   (a) Aluminum   2.2   – 3.2   g/mL
   (b) Brass   8.3   – 9.3   g/mL
   (c) Steel   7.3   – 8.3   g/mL

Waste requirements: none

EXPERIMENT 3: Preparation and Properties of Oxygen

Since this is the first experiment involving the collection of a gas by water displacement, the use of a pneumatic trough, including the handling of inverted bottles of water, should be demonstrated.

This is also the first experiment in which the students are asked to write chemical equations. It therefore, affords an opportunity to explain and discuss the concept of chemical equations during the introductory discussion period. Refer students to Study Aid 2.

Safety Precautions:

1. Hydrogen Peroxide:
   a. Nine percent hydrogen peroxide may be prepared from commercially available 30 percent H_2O_2. (Purchase the peroxide with added preservative.)
   b. Use gloves and eye protection when handling 30 percent H_2O_2, since it is a hazardous chemical.
   c. Students should not be allowed to handle 30 percent H_2O_2, but may safely handle the 9 percent solution. However, any skin area coming in contact with the 9 percent solution should be washed with water immediately.

2. Students sometimes have trouble getting the steel wool to burn in oxygen. Emphasize that the steel wool must be quickly transferred from the flame to the bottle of oxygen and, in fact, must be glowing when it is inserted.

3. Students should be reminded not to look directly at the flame produced by burning magnesium.

4. The Büchner funnel-vacuum flask apparatus for the separation of the MnO_2 in the generator flask from the H_2O_2 mixture should be demonstrated before students begin the experiment. One of these setups at a sink that is accessible to everyone in the class is probably adequate for 24 students who work in pairs. The filter should be monitored and changed as needed when it is being used by many students. The level of the reacted H_2O_2 mixture in the vacuum flask should also be monitored and emptied into the sink as needed.

5. Instructor Demonstrations: The spontaneous combustion that occurs when water is added to sodium peroxide on cotton makes a spectacular demonstration but should be conducted with caution. Test the sodium peroxide in advance to make sure it is still active.

Stability of Hydrogen Peroxide Solution
Thirty percent H_2O_2 should be stored in the refrigerator. Nine percent H_2O_2 is stable for long periods of time when stored in the refrigerator but decomposes at room temperature. Store the 9 percent solution in brown bottles. If the 9 percent solution has been standing at room temperature for several days it may be necessary to use more than 50 ml to obtain the required amount of oxygen. It is advisable to prepare the 9 percent solution shortly before use.

Waste requirements:
A bottle for recycling unreacted 9 percent H_2O_2 should be provided, the Büchner funnel-vacuum flask should be setup for disposal of MnO_2.

EXPERIMENT 4: Preparation and Properties of Hydrogen

In addition to illustrating some of the properties of hydrogen gas, this experiment provides qualitative experience with pH and the reactivities and reaction rates of different acids with metals. Additional experience in handling simple chemical equations is also provided.

If Experiment 3 has not been done, refer the students to it for directions on using a pneumatic trough and gas collecting bottles. Also refer the students to Study Aid 2.

A great deal of student time can be saved by preparing a sample bottle containing about 10 grams of mossy zinc to help students approximate this amount. Since the unused zinc is collected for reuse, it doesn't matter if more than 10 grams is used.

It is important to emphasize the distinction between combustible substances and substances which support combustion.

Because of the great tendency of hydrogen gas to escape, success with the tests of Part D is dependent on lifting the bottle straight up from the table top **without** tilting it to one side.

Safety Precautions:

1. It is advisable for the instructor to dispense one cube of sodium metal to each student. These cubes should be no larger than 4 mm on an edge.
2. Remind students not to look directly into the test tube when sodium and water are reacting. The escaping hydrogen may ignite and splatter unreacted sodium metal and sodium hydroxide solution.
3. Remind the students to keep lighted burners and all other flames away from the hydrogen generator.

Waste requirements:

(1) A container for unreacted metal strips is needed, and (2) a recycling jar for used zinc (rinsed with water) should be provided.

FIRST GRAPHING SESSION

Early in the semester before graphing skills are required for experimental data, we devote a full laboratory session to graphing, both by hand and by computer. Since Experiment 5, Calorimetry and Specific Heat involves a graph, the graphing session is usually done before Experiment 5. You may choose to do this in a different sequence or have students learn this on their own. During this session Study Aid 3 (SA3) in its entirety is completed:

1. Plotting two variables by using graph paper; two practice graphs are included in SA3.
2. Completion of Exercise 9 on plotting data and reading graphs.
3. Completion of a computer graph of data in SA3 using Excel, Chart Option (within Microsoft Office 2000 or XP.) If your facilities lack computers in the laboratory, students can complete this component in a computer lab or at home; the Chart Option can also be accessed through MS Word. The directions can be applied to both Windows-based PC computers and MacIntosh. System requirements are described in the Study Aid.

At first the computer graphing requirements may be difficult for some students because they have little or no experience with computers. These students need some help and support as they learn how to manipulate a mouse and gain confidence with the keyboard. They benefit from working with more sophisticated students and are usually transformed by successful completion of their first graph.

EXPERIMENT 5: Calorimetry and Specific Heat

This experiment provides support and reinforcement to several important areas:

1. determination of specific heat of a known substance by using a calorimeter;
2. completion of two trials of a procedure to determine the precision of the specific heat measurements;
3. comparison of the experimental specific heat to a theoretical value to determine the accuracy of the specific heat measurements;
4. practice of graphing skills using atomic mass vs. specific heat data provided in the discussion.

We use the same metal slugs that were used for the density measurements in Experiment 2. Sometimes (on a very dry day) there is enough static electricity generated by the styrofoam cups that the electronic balances are erratic. It is good to have a low-tech balance available. Common sources of error include:

1. test tube in the boiling water should not rest on the bottom of the beaker.
2. boiling water should not evaporate below the top of the metal slug in the test tube.
3. temperature of the thermometer in the boiler should not be greater than 100°C.
4. transfer of hot metal to the calorimeter must be done quickly as heat is lost to the surroundings very quickly. After the transfer of the metal into the calorimeter, the water should be stirred while monitoring the temperature.
5. the volume of water in the calorimeter should be just enough to cover the metal sample. Too much water results in a very small temperature change because of the high water mass absorbing the heat.

For Trial 2, the water should be colder than Trial 1 to illustrate that the temperature change experienced by the water in the calorimeter, not the initial temperature, is the essential quantity in determining the heat absorbed by the water. A one-liter bottle of distilled water in an ice bath will provide enough water at 5–10°C for an entire class.

The discussion provides a sample calculation as a template for the experimental calculations and we insist on students showing setups on the report form as well as answers.

Evaluation:
High accuracy (i.e., agreement with the theoretical value for the metal sample) is not a priority in the experiment. We look for an improvement in accuracy from the first trial to the second trial. Graphing skills, problem setups, and insight into heat concepts are the priorities here.

Waste requirements: none

EXPERIMENT 6: Freezing Points—Graphing of Data

This experiment has several objectives:

1. To provide practice in taking and plotting data using graphing skills learned in Study Aid 3. Graphs are done by hand or by computer.

2. To determine the experimental freezing point of a pure substance.

3. To measure the freezing point depression due to the addition of a solute to the pure substance.

4. To observe the phenomenon of supercooling and the effect of stirring a liquid on both supercooling and the freezing point.

Acetic acid is convenient to use because it can be solidified using ice-water and is generally available. Since acetic acid is hygroscopic, fresh glacial acetic acid should be used in this experiment. Slotted corks or stoppers are used in order to see the entire thermometer scale.

In order to compare the three trials, students should take care to make sure the amount of ice/water in the water bath remains constant for each trial. Supercooling will not usually occur if the liquid is stirred. If carefully left **undisturbed,** the temperature of the liquid usually goes well below the expected freezing point of pure acetic acid, (definite supercooling). Then, the thermometer is moved and rapid crystallization should occur. Exceptions to this usually mean that students have disturbed the system inadvertently.

Evaluation:

The primary purpose of the experiment is to provide experience in collecting and plotting experimental data. Questions on the report form direct students to the analysis of the data from their graphs. Grading is based primarily on the precision and care exercised in taking data and plotting this data on paper.

Waste requirements:

A bottle should be provided to collect the acetic/benzoic acid mixture which is left at the end of the experiment. Students should be reminded NOT to return this mixture to the original stock bottle of glacial acetic acid.

EXPERIMENT 7: Water in Hydrates

This experiment is divided into a qualitative part (A) and a quantitative part (B). The latter requires a considerable amount of weighing time. If the supply of balances is limited, it is advisable to divide the class into two groups and have one group begin on Part B while the other group is doing Part A.

The technique of heating to constant weight is introduced for the first time in the quantitative part of this experiment and should be emphasized. The technique of using crucible tongs should be demonstrated.

Although the title of this experiment is "Water in Hydrates", some information on other types of thermal decompositions may be included in the introductory discussion period.

Unknowns:

1. Magnesium sulfate heptahydrate ($MgSO_4 \cdot 7H_2O$)
2. Zinc sulfate heptahydrate ($ZnSO_4 \cdot 7H_2O$)
3. Barium chloride dihydrate ($BaCl_2 \cdot 2H_2O$)
4. Strontium chloride hexahydrate ($SrCl_2 \cdot 6H_2O$)
5. Sodium bicarbonate ($NaHCO_3$)
6. Potassium bicarbonate ($KHCO_3$)

Sodium bicarbonate and potassium bicarbonate are readily available and are suitable as unknowns for this experiment. They are quantitatively decomposed to the respective carbonates by heating under the prescribed conditions. It is not essential for the student to know whether the sample is a true hydrate or one of these substances. The mass loss is calculated as water.

We issue the unknowns in 3×4.5 inch coin envelopes. Inexpensive plastic vials are also available for this purpose. Unknowns should not be stored in paper envelopes for long periods of time as they tend to lose water of hydration. We issue 6 to 7 g of unknown to each student.

Evaluation:

We weight the value obtained on the unknown at 40% of the total points assigned to the experiment using the following scale:

Deviation* (± %)	Points off
0 – 1	0
1 – 1.5	5
1.5 – 2	10
2 – 3	15
3 – 4	20
4 – 5	25
5 – 6	30
6 – 8	35
> 8	40

*From these accepted values for percent water:

$MgSO_4 \cdot 7\,H_2O$	51.2%
$ZnSO_4 \cdot 7\,H_2O$	43.8%
$BaCl_2 \cdot 2\,H_2O$	14.8%
$SrCl_2 \cdot 6\,H_2O$	40.5%
$NaHCO_3$	36.9%
$KHCO_3$	31.0%

Waste requirements:

A container for the solid residues remaining after the decomposition of the known and unknown hydrates is needed. (Some will contain heavy metals: Cu^{2+} Zn^{2+}, Sr^{2+}, Ba^{2+}.) The unused portions of the unknowns are returned to the instructor.

EXPERIMENT 8: Properties of Solutions

The experimental portion of this experiment is relatively long. Consequently some students may not be able to finish both the experimental work and the report form in the three-hour laboratory period. We observe how the experiment is proceeding with regard to time and, if necessary, tell the students (near the end of the period) to complete the experimental work, finish the report form after class, and hand it in at the beginning of the next laboratory period.

In the interest of time economy it is a good idea to remind the students to proceed with other parts of the experimental work while the potassium chloride solution is evaporating to dryness (Part A).

It is advisable to prepare the saturated potassium chloride solution at least a day in advance in order to allow sufficient time for it to attain temperature equilibrium in the laboratory.

Molarity is included in the common ways to express concentration of a solution although no work pertaining to molarity is included in the experiment.

Evaluation:

Part A. Although the actual results obtained in this part are not heavily weighted, we especially check the data to make sure that the masses are expressed to the precision stated in the experimental directions (to the highest precision of the balance).

Waste requirements:
Three bottles should be provided: (1) Waste Organic Solvents (Decane); (2) Waste Kerosene mixtures; and (3) Waste Heavy Metals (Ba^{2+}).

EXPERIMENT 9: Composition of Potassium Chlorate

Good budgeting of time is important in this experiment. When two samples are analyzed, the heating and weighing procedures should be overlapped. For example, analysis of the second sample should be started while the first sample is being heated; the second sample should then be heated while the first is being cooled for weighing; etc.

If the student has only one crucible and cover, there is not enough time to do more than one sample.

If the crucibles are perfectly clean, 20 to 30 minutes can be saved with only a little loss of accuracy by omitting the initial heating and cooling of the empty crucibles (Part A).

Emphasize that the initial heating of the $KClO_3$ must be gentle to avoid losses by frothing.

Safety Precautions:
Potassium chlorate has potential for fire and explosion hazards and the following are important to keep in mind.

1. Use reagent grade $KClO_3$.
2. An explosion may occur if molten $KClO_3$ is allowed to come into contact with a rubber stopper or other organic materials.
3. Fires may result if $KClO_3$ is mixed with combustibles such as waste paper.

Evaluation:
We evaluate the percentage composition in Part A.4 using the student's best result (nearest to 39.15% oxygen). A maximum of 30 points is given to Part A.4 according to the following scale.

Percentage Oxygen (A.4)

Deviation from 39.15%

Deviation (± %)	Point deduction
0 – 2	0
2 – 3	5
3 – 4	10
4 – 5	20
> 5	30

The remaining 70% is distributed over the other parts of the experiment. Since any experimental error in Part A.4 leads to a corresponding error in Part A.5, we do not deduct additional points in A.5 unless the student has made an additional error.

Waste requirements:
A bottle should be provided for Waste Heavy Metals (Ag^+).

EXPERIMENT 10: Double Displacement Reactions

This experiment is fairly easy to perform and requires appreciably less time than most of the others that are included in this manual. If a relatively short discussion period is used, Experiment 10 and Experiment 11 can be done in a single laboratory period—or a laboratory examination may be scheduled. If Experiments 10 and 11 are done on the same lab day we suggest that half of the class start on Experiment 10 and half on Experiment 11 in order to avoid crowding for reagents.

Alternatively, a rather detailed discussion on writing and balancing chemical equations may be presented since the students are asked to write a considerable number of equations. The first page of Exercise 7 is suggested as a follow-up for this experiment.

It is important to emphasize that only approximate volumes of solution are needed. Otherwise undue amounts of time are consumed in measuring the volumes used.

The Solubility Table in Appendix 5 should be pointed out and its use discussed and illustrated—for example, is either of the products in a double displacement reaction a precipitate?

Waste requirements:
A bottle for Waste Heavy Metals (Ag^+, Ba^{2+}, Cu^{2+}, Zn^{2+}) should be available.

EXPERIMENT 11: Single Displacement Reactions

This is another short experiment; and for the most part, the remarks made under Experiment 10 apply equally well to this experiment. The experiment enhances the students' knowledge of relative reactivity of metals, and gives them further practice in equation writing. Metal strips must be cleaned before use to obtain satisfactory reactions.

A suggested demonstration for the introductory discussion period is simply to dip a clean iron nail into a copper(II) sulfate solution. Visible evidence of the plating out of metallic copper on the nail is observed almost immediately.

Waste requirements:
A bottle for Waste Heavy Metals (Ag^+, Cu^{2+}, Pb^{2+}, Zn^{2+}) should be available.

EXPERIMENT 12: Ionization—Electrolytes and pH

The ionization demonstration and discussion probably will consume about 45 minutes. During the demonstration the students should complete the table in Part A.

The conductivity apparatus we use is illustrated in figure 12.1 in the laboratory manual. It consists of a support base, No. 12 covered copper wire, a ceramic light socket, a clear glass 40 watt bulb (utility bulb), and an extension cord. The support base can be made from pressed fiber board, plywood, or various plastic materials. This apparatus allows demonstrations to be conducted with test tube quantities of reagents. In addition, a demonstration set of solutions (in test tubes) can be stored and used over and over again. The concentration of the salt solutions tested is about 0.1 M.

In Part A8 of the demonstration it is advisable to use only one drop of dilute sulfuric acid, otherwise the volume of barium hydroxide solution and the time required to add it become excessive.

If practical to the schedule, this demonstration can be shown in the lecture section, thereby avoiding repeating it several times in smaller laboratory sections.

Since there are a number of parts to this experiment, the slow or overly meticulous student may have difficulty completing it in a three-hour laboratory period.

We set up at least two pH meter stations in the laboratory. At one of these stations, we set out labeled beakers of the three dilutions of hydrochloric acid in order from the most dilute (0.001 M) to the least dilute (0.1 M). At the second of these stations, students bring their own beakers of alkaline solutions for pH measurements. At the alkaline station, a beaker of dilute acetic acid is provided to neutralize residual base on the electrodes before rinsing with distilled water. For larger classes, additional pH stations can be set up if more meters are available.

Waste requirements:
Since students will not be handling any of the solutions involved in the demonstration, the heavy metal ions (Ba^{2+}, Ni^{2+}, Cu^{2+}) will not require a special waste bottle for student use.

EXPERIMENT 13: Identification of Selected Anions

This experiment introduces qualitative analytical procedures. Arsenate has been eliminated from this experiment in this edition. In addition to testing the reaction characteristics of six different anions, each student is required to identify two unknowns chosen from among the six.

This experiment also introduces the concept of un-ionized, total ionic, and net ionic equations. If the instructor wishes to stress this strongly, he or she might choose to illustrate this concept with equations from Experiment 10.

Make sure that the chlorine water has been freshly prepared—or that its color is yellowish green.

Unknowns:
Our system for dispensing unknowns is as follows: When students are ready to begin work, they each bring two clean test tubes to the instructor. The instructor gives the student approximately 10 mL samples of the unknowns, together with the corresponding numbers. Dispensing of the unknowns can, of course, be handled equally well by stockroom personnel.

The following is a suggested listing of unknowns. We use eight bottles of solutions to keep the students from discovering "the system". Our order is Br^-, Cl^-, SO_4^{2-}, Br^-, PO_4^{3-}, I^-, CO_3^{2-}, PO_4^{3-}. The bottles of unknowns are filled from the stock solutions of the sodium salts used in the experiment.

Evaluation:
We assign 25 points for each of the two unknowns; the other 50 points are distributed over the balance of the experiment.

Waste requirements:
Two waste bottles are needed: (1) Waste Organic Solvents (decane); and (2) Waste Heavy Metals (Ag^+, Ba^{2+}).

EXPERIMENT 14: Properties of Lead(II), Silver, and Mercury(I) Ions

This experiment covers the usual qualitative analysis of the silver group cations. It includes the chemical reactions of the individual ions, the examination of a known solution containing all three cations, and the analysis of an unknown containing one, two, or three of the cations. If Experiment 13 has not been done, the student should be referred to that experiment for a discussion of un-ionized, total ionic, and net ionic equations. NOTE: The use of chromates has been eliminated and NaI is used to confirm Pb^{2+}.

We find the students are often disturbed by negative results on their unknowns. They need to be reassured that for an unknown, a negative result is just as significant to the analysis as a positive result. For instance, when addition of sodium iodide does not produce a yellow precipitate, Pb^{2+} is absent.

Unknowns:
The following suggested list of unknowns covers all possible combinations using the three cations of the group.

Ag^+ (stock solutions)

Ag^+, Hg_2^{2+}
Ag^+, Pb^{2+}
Ag^+, Pb^{2+}, Hg_2^{2+} (known solution)
Pb^{2+} (stock solution)
Hg_2^{2+} (stock solution)
Pb^{2+}, Hg_2^{2+}

Preparation of the Known and Unknown Solutions:
The known solution should contain 28 g of $HgNO_3 \cdot H_2O$, 17 g $AgNO_3$, and 33 g $Pb(NO_3)_2$ per liter of solution. Prepare the mixture as follows: mix $HgNO_3 \cdot H_2O$ with 50 mL of conc. HNO_3 and 200 mL of water. While stirring the mixture, gradually add about 650 mL of additional water and continue stirring until all of the salt is dissolved. Then add the $AgNO_3$ and $Pb(NO_3)_2$, stir until the solids are dissolved, and dilute to 1 liter.

The unknown solutions may be prepared in a similar manner using the same amounts of the salts needed for each unknown. The solutions will be 0.10 M in each cation. Stock solutions used in the first part of the experiment are satisfactory for unknowns containing only one kind of cation.

We issue 5 to 10 mL of unknown to each student using the method of dispensing unknowns given in Experiment 13.

Evaluation:
We assign 30 points for the unknown and distribute the other 70 points over the balance of the experiment.

Waste requirements:
A bottle for Waste Heavy Metals (Pb^{2+}, Ag^+, Hg^{2+}) is needed. To eliminate constant trips to this waste bottle, students can pour these solutions into a beaker at their stations and empty the whole beaker into the Waste Heavy Metals bottle at the end of the experiment.

EXPERIMENT 15: Quantitative Preparation of Potassium Chloride

This experiment makes practical use of the principles involved in stoichiometric mass to mass calculations using the mole method. Study Aids 4 and 5, and Exercises 10 and 11 are useful supplements for this experiment.

It saves time if students set up and start heating the water bath before they start weighing the sample. It is important to follow instructions for drying the KCl carefully. To avoid splattering in the second stage of heating (without the water bath) it may help to move the burner back and forth by hand under the dish rather than heat steadily.

Sometimes, especially when the lab is humid, heating to constant weight by the usual standard (0.05 g) does not occur after two heatings, even when students are very meticulous. Therefore, we have relaxed the standard to a difference of 0.08 g between the first and second heatings. If a third heating is necessary and constant weight within 0.08 g is not obtained, we accept the final weight.

Some points the instructor may need to discuss during the prelab session or as the students work are:

1. It is useful to go through a sample calculation to emphasize the importance of using an excess amount of the 6.0 M HCl and how that correlates with the gas bubbles observed.

Students may not yet have any experience using molarity in a dimensional analysis setup and should be referred to Study Aid 5.

2. The importance of heating to constant weight should be emphasized, especially if students have not used this technique previously (it was introduced in Experiments 7 and 9).

3. So there will be sufficient excess of 6.0 M HCI, the starting sample of $KHCO_3$ must not exceed 3.0 grams.

Waste requirements: none

EXPERIMENT 16: Electromagnetic Energy and Spectroscopy

During the prelab we give students an overview of the three parts and demonstrate the generation of waves with a stretched spring. They work in groups of 4–5 for Part A (wave properties) and individually with their own hand-held spectroscope for Part B (emission spectra). Because we have three spectrophotometers in the laboratory, they work in three teams for Part C. If it is a very large class and the teams are too large then half the class works on Part C while the other half completes Part B.

The students go out to the hall where there is plenty of room to extend and vibrate the springs needed to generate the waves in Part A. Two students sit on the floor about 4 meters apart and stretch the spring between them. *The spring never leaves the floor.* Students are cautioned to hang on to the spring when it is stretched so it won't recoil and snap back at the person on the other end. Another student is designated the "timer" with a stopwatch or watch with a second hand. The counting must be done out loud. We tell the students to watch a place on the floor that a wave peak touches and to count every time the spring hits that spot. The rest of the team evaluates the wave forms and tells the two on the spring when the frequency is correct and when the timing and counting should be started. The spring does most of the work and it always works very well. Students return to the lab for data analysis. The graph of wavelength vs. frequency is done by hand using skills learned in Study Aid 3.

For Part B (emission spectra) we use the ceiling lights in the lab for the fluorescent light. We set up several incandescent lamps, 2 hydrogen vapor lamps and 2 neon vapor lamps around the laboratory. It is important for the instructor to circulate and check the student spectrum sketches to make sure they are looking at and sketching the correct light. Frequent trips to the spectra chart on the wall are encouraged. The difference between energy content and wavelength is constantly reviewed as they complete the hydrogen electron transition diagram on the report form. If the light in the room is too bright, the hydrogen and neon spectra are confused with stray light. If this is a problem, turn off the room lights.

For Part C (absorption spectra) we turn on the Spectronic 20s when the class starts Part B. As soon as a few students finish B, we demonstrate the operation of the instrument and continue with the group demonstrations until all the instruments are busy. Then, as more students finish B, they join the ongoing groups and help each other. Every student is required to take a few measurements. By sharing data, this goes fairly quickly and everyone gets finished. Trips to the spectra chart to match wavelengths transmitted and absorbed with color help students relate the absorption data to the color of the solutions.

The graph of the absorption spectra can be done on the computer or can be done by hand. If students are running late and do not have time for this last graph, they are allowed to finish it on their own time and turn in the report form with the graph at the next lecture.

Evaluation:
Assigning points to each part of the report is each instructor's prerogative.

Waste requirements:
A bottle for Heavy Metal Wastes (Ni^{2+} and MnO_4^-) is needed.

EXPERIMENT 17: Lewis Structures and Molecular Models

Students work in pairs so a class of 24 students requires 12 model kits. The number of kits therefore, will depend on how many labs are run simultaneously. The models which work best are the ball and stick or "open" type.

At first students might seem overwhelmed by the expectation of completing all 16 molecules/ions within a three hour lab period. However, we have found that this is a very reasonable expectation. Because the instructor is circulating to look at the models, students can get some assistance with the Lewis structures where necessary. The last column on the chart (Dipole vs nonpolar) causes the most confusion and students need repeated clarification on the difference between bond polarity and the distribution of the polar bonds within a molecular species.

The experiment works best if the concepts of Lewis structures, molecular geometry, bond angles and polarity have been covered in lecture. However, by giving each pair of students a kit and working through the discussion as a class, the concepts can be taught during an extended prelab. Even after exposure in lecture and an extended prelab, there is much mental confusion until students start hands-on working through the examples in the procedure. By the end of this experiment, students are usually pleasantly surprised with their understanding of Lewis structures and molecular geometry.

Students must use a sharp **pencil** to complete the report form. Extra copies of the report form are available for those who do not take the instructions about the sharp pencil seriously until after they have worked a few examples and erased holes in their paper. Students should be told to work each example all the way across a row rather than building models as fast as possible.

The instructor should circulate from group to group checking models so that students can dismantle the checked models and reuse the components. Column C on the report form is checked for both partners when a model is approved.

When students get to the structures with two central atoms, ex. C_2H_6, there may be some trouble with the geometry and bond angles until they realize that the carbon atoms serve as central atoms and as peripheral atoms for each other. Most see this very quickly with the model in hand once the peripheral atoms on one end are covered up.

Evaluation:
There are 16 molecules/ions on the report form. Each requires 7 student tasks so there are a total of 98 points that can be earned by completing all the examples. Angles on the Lewis structures do not have to be drawn accurately but all the dots and bonds must be correct. Column C is graded during the lab session. All the other columns are graded after the papers are turned in. The Questions and Problems provide another possible 15 points. The number of points earned is divided by 113 to determine the percent grade.

There are two examples for which more than one possible Lewis structure can be drawn. Since resonance and isomerism are beyond the scope of this experiment, we accept either structure. For $C_2H_2C_{12}$, the molecular polarity depends on the isomer drawn and it is important to cross reference these when the report is graded.

Waste requirements: none

EXPERIMENT 18: Boyle's Law

The kits for this experiment to demonstrate Boyle's Law can be ordered from several vendors. The kits include everything, even the silicone grease, except the applied masses and the vernier calipers. We use the Cenco kits so references in the instructions to color may be different for kits from a different vendor. The masses shown in the diagrams are heavy slotted masses in 0.5 kg and 1 kg sizes. With these kits, the tip of the syringe barrel extends beyond the surface of the weight platform so the first applied mass must be a slotted weight (so the syringe tip can be inside the slot) allowing the mass to lie flat on the platform. If there are not enough slotted weights available then after the first applied mass, bricks work very well. Combinations of bricks and slotted weights work well also.

In our experience the stated weight of these masses is accurate to two significant digits so to get the third significant digit it is necessary to weigh each mass if possible. A single electronic digital balance per laboratory with a large capacity (1200 g–2500 g) is enough for students to share. If this is not available, the masses and bricks can be preweighed and labeled with tape.

A vernier caliper is necessary to get enough significant figures for the measurements of the inside diameter of the syringe barrel. We use inexpensive plastic calipers. Most students are not familiar with the use of a vernier caliper so it will be necessary to explain its use.

An interesting variation on the procedure is to have each student group start with a different initial volume (rather than at 35 cm^3 as stated in the instructions).

We have students work in groups of 2–3 per kit. It is very important for someone to stabilize the syringe as the masses are added so the syringe does not tip over or break. It is also possible to stabilize the syring by attaching it to a ring stand with a clamp loosely around the barrel.

The graph can be drawn by hand or computer.

Evaluation:

Recording data and calculation set ups	40%
Questions and Problems	30%
Graph	30%

Waste requirements: none

EXPERIMENT 19: Charles' Law

To obtain reproducible results in duplicate runs it is essential that both the flask and stopper assembly be dry. We find that reproducible results are more readily obtained when the flask is flame dried at the beginning of each run. However, in order to avoid cracking the flask, it should be dried with a towel before heating in the flame.

The interpretation of the data is a bit complex for many students, but with a little patience and guidance they eventually work it out. The calculations include a correction for water vapor in addition to the Charles' law calculation.

Evaluation:
We consider that 10 percent agreement between the calculated and experimental values (No. 3 under Calculations) is reasonable for this experiment. The straight line on the graph should intersect the temperature axis within ±25° of −273°C.

Waste requirements: none

EXPERIMENT 20: Liquids—Vapor Pressure and Boiling Point

The three liquids for Part A should be placed in 125 mL Erlenmeyer flasks and stored on the reagent shelf for students to use to moisten their filter paper-covered thermometer bulbs.

Part B is best done by two students working together, especially since each apparatus requires two 25 × 200 mm test tubes and two thermometers. Boiling chips need to be supplied. **Students should be cautioned that failure to follow instructions about scrubbing the test tubes with detergent and rinsing will make it difficult to read the thermometer and collect the data needed.**

In Part B, a number of operations must be done in rapid and proper sequence at the time heating the test tubes is stopped. If heating is stopped too soon before the screw clamp is closed, water backs up into the tube. If heating is continued after the clamp is closed, the rubber stopper may pop out of the test tube. A review of these operations and the importance of the timing is of value. If the water in the closed tube stops boiling soon after heating is discontinued, there is probably a leak in the system.

Waste requirements: none

EXPERIMENT 21: Molar Volume of a Gas

This experiment integrates the variables of the gas laws and related stoichiometry. It is scheduled late in the semester when measurement skills are better developed. Students often get very excited when an experimental volume close to 22.4 L/mol pops out of the calculator. We have students work in pairs.

The syringe-stopper assembly should be prepared for the students and, for safety and security reasons, should be checked in and out. An additional safety feature is to snip off the end of the needle with a wire cutter. Disposable syringes and needles can be purchased individually or in bulk. They can be used over and over again so there is no need to keep a large stock on hand. The needles need to be heavy enough to push through a rubber stopper without bending and long enough to extend all the way through.

An alternate syringe-stopper assembly which eliminates the use of a needle is possible for those who prefer to avoid the needle-stopper assembly. The alternate assembly uses a liquid transfer pipetter (Fisher CLS30466-2) and requires a more complicated insertion through the stopper. We have used the needlestopper assembly described for many years with no problems whatsoever. Students may need a little practice reading the volume in the syringe.

3.0% hydrogen peroxide is purchased locally from a drug store and stored in the refrigerator. We have used hydrogen peroxide stored for a least a year and had no problems with the concentration. Manganese dioxide must be powdered, not granular.

A big problem in this experiment is inverting the filled graduate into the beaker of water without letting any water out (and air bubbles in). The pouring spout on most graduates are difficult to seal because a thumb is too small. If the whole hand is cupped over it the beaker opening may not accommodate the whole hand. We solve this by using 2 L beakers or battery jars. Graduated cylinders without pouring spouts will also help. If all else fails, the volume of air admitted to the cylinder during this inversion process can be compensated for in the calculations or considered part of experimental error.

Small errors in reading the volume of the gas generated will not increase the percent error as much as small errors in reading the volume of peroxide used.

Evaluation:
Percent error of 5% or less is very reasonable. This translates into 22.4±1.1 (or a range of 21.3 L/mol to 23.6 L/mol). Since the significant figures are limited to 2 by the concentration of the peroxide solution any answer between 21 and 24 is acceptable.

Our emphasis is on correct application of the gas laws, setting up the calculations, the chemical reaction and its stoichiometry.

Waste requirements:
A Büchner funnel-vacuum flask should be setup for disposal of MnO_2. See Experiment 3 for a diagram of this setup.

EXPERIMENT 22: Neutralization—Titration I

In this experiment each student will need a buret (50 mL or 25 mL). The techniques of filling, dispensing liquid, and reading the buret are demonstrated.

We dispense KHP in small stoppered test tubes (13 × 100 mm). We weigh 4 grams of KHP into one tube and fill the additional tubes needed to about the same level. Each student picks up a tube of KHP from a rack, uses what he or she needs, and then returns the tube to the rack. Thus, the tubes can be stored, refilled, and used again.

It is worthwhile to point out to the students why the flask that receives the base solution must be perfectly dry–and why the flask that receives the KHP need not be perfectly dry. The use of a wash bottle can be demonstrated to rinse the flask with distilled water, to wash down any KHP that sticks to the walls of the flask, and to wash the walls of the flask during the titration.

Significant figures prove to be troublesome in this experiment. Students tend to round off the moles of KHP to 0.005. Students also need to be reminded that KHP is an abbreviation for potassium hydrogen phthalate and is not a compound of potassium, hydrogen, and phosphorus.

In an experiment such as this where good precision is possible, it should be made clear to the students that they should not settle for badly missed end-points, as indicated by bright pink instead of a faint pink solution at the end of the titration.

Unknowns:

Students will need about 250 ml, of unknown base if they will be doing Experiment 23 also, or about 100 mL if they will do only Experiment 22. We dispense our base "unknowns" as follows: students bring a **clean, dry** 250 mL Erlenmeyer flask to the stockroom and exchange it for a numbered flask containing the base sample. An alternate procedure is to place three or more large reagent bottles containing different concentrations of base solution in the laboratory and assign each student one of these bottles. The first procedure saves considerable base, for there is much waste when the supply of unknown is left out in the laboratory. Sample flasks are filled in advance by stockroom personnel.

For the suggested range of KHP masses, base molarities should be in the range of 0.30 to 0.35 M. We prepare these NaOH solutions in 20 liter carboys and then standardize against KHP as in the experiment. The solutions are made using NaOH pellets. A typical set of unknowns is:

Approximate molarity	Grams NaOH per liter	Grams NaOH per 20 liters
0.304	12.15	243
0.314	12.55	251
0.323	12.90	258
0.333	13.30	266

Evaluation:

We evaluate the unknowns on absolute deviation from the known molarities. Thirty-five (35) points are allowed for this determination (Calculation No. 4). The other 65 points are distributed over the balance of the experiment. The scale we use for evaluating the unknowns follows:

Deviation (±)	Points off
0.000 − 0.003	0
0.003 − 0.006	5
0.006 − 0.009	10
0.009 − 0.012	15
0.012 − 0.015	20
> 0.015	35

Waste requirements: none

EXPERIMENT 23: Neutralization—Titration II

Each student will need a buret and a 10 mL volumetric pipet. It is advisable to demonstrate the technique of using a volumetric pipet. If Experiment 22 has not been done, the technique of filling, dispensing liquid from, reading, and eliminating air from the tip of the buret should be demonstrated.

Each student will require about 150 mL, of NaOH solution prepared and standardized in the concentration range of 0.30 to 0.35 M. See Notes on Experiment 22 for suggestions on the preparation and standardization of this solution in bulk quantities. **Remind students not to pipet by mouth.**

Students should have enough NaOH solution left over from Experiment 22 to do Experiment 23. However, you will need to tell them the molarity of their unknown base used in Experiment 22 in order for them to do the calculations in Experiment 23.

Unknowns:

A 50 mL sample of unknown acid solution is needed for each student. Since we have several hundred students in our laboratories each year, we prepare these acid solutions in 20 liter carboys. Seven different solutions are prepared by diluting sulfuric acid to the approximate concentrations shown in the following table. We standardize these acid solutions by titrating them with one of the base solutions used in Experiment 22.

Approximate H+ Molarity	Milliliters concentrated sulfuric acid* per liter
0.48	13.35
0.50	13.90
0.52	14.45
0.54	15.00
0.56	15.55
0.58	16.10
0.60	16.65

*Based upon 96% sulfuric acid having a density of 1.84 g/mL.

Samples are dispensed by having each student take a **clean, dry** 125 mL Erlenmeyer flask to the stockroom and exchange it for a numbered flask containing the sample. Sample flasks are filled in advance by stockroom personnel.

Students are asked to calculate the concentrations of their unknowns as molarities of hydrochloric acid. However, we use sulfuric acid to prepare the unknowns because it gives more stable solutions than hydrochloric acid. For the sake of simplicity we have chosen not to introduce normality or diprotic acids into the experiment.

About 35 mL of vinegar are required for each student. We buy and use commercial white vinegar. "Synthetic vinegar" can be made by simply diluting glacial acetic acid with water to form a solution containing four to five percent acetic acid.

Evaluation:

We evaluate the acid unknowns on absolute deviation from known molarities. Thirty-five (35) points are allowed for this determination (Part A.4). The scale we use for evaluation follows:

Deviation (±)	Points off
0.000 − 0.003	0
0.003 − 0.006	5
0.006 − 0.009	10
0.009 − 0.012	15
0.012 − 0.015	20
> 0.015	35

We allow 15 points for mass percent acetic acid in the vinegar (Part B.6). The students must be within 0.30% of the correct value as follows:

3.70 − 4.30% (4% vinegar)
4.70 − 5.30% (5% vinegar)

The other 50 points are distributed over the balance of the experiment, emphasizing mathematical set-ups and significant figures.

Waste requirements: none

EXPERIMENT 24: Chemical Equilibrium—Reversible Reactions

Few, if any, experimental difficulties are encountered when the students perform this experiment. However, many students have diffculty interpreting the equilibrium concepts that are involved.

The equilibrium equations are given for all six reversible reactions studied. Students should be directed to refer to these equations when answering the questions on the report form.

The evidence of an equilibrium shift in Parts A and B is the formation of salt crystals; the formation of crystals is dependent on the solutions being saturated. The shifts of equilibrium in the other parts (C, D, E, and F) are detected by color changes.

The discussion period may well emphasize the equilibria in Parts C and E as we have found that students in general seem to have some difficulty interpreting the results of these parts. In Part C.4 they seem to miss the fact that the shift of equilibrium is due to the removal of SCN^- ion by precipitation as AgSCN. In Part E.2 they seem to miss the fact that the equilibrium shift is due to the increased chloride ion concentration caused by the greater solubility of ammonium chloride at elevated temperatures.

Waste requirements:
A bottle for Waste Heavy Metals (Ag^+, Fe^{3+}, Cu^{2+}, Co^{2+})

EXPERIMENT 25: Heat of Reaction

This experiment can be scheduled at any point in the sequence following Experiment 13 in which the concept of net ionic reactions is introduced. The work of this experiment is designed to reinforce the concept that energy changes are associated with all types of chemical reactions.

In Parts A and B, temperature changes are measured to demonstrate the difference between heat of hydration and lattice energy. In Part C, the heat of neutralization for a strong acid plus a strong base, is actually calculated using the temperature change observed and the calorimetry equation.

Students need to be reminded that all measurements (concentration, temperature, and volume) need to be taken to at least two significant figures in order to obtain a final value for the heat of reaction with two significant figures.

Calorimeters made from inexpensive Styrofoam cups are surprisingly effective in minimzing heat losses. This contributes to the reproducibility of temperature changes in duplicate trials which in turn contributes to the agreement of the experimental heat of neutralization with the theoretical heat of neutralization (55.9 kg/mol). The theoretical heat of neutralization was obtained from the Heats of Formation (from any H_f reference table) for H_2O (1) from H^+ and OH^- as follows:

$$\begin{array}{ccccc} 0 & & -229.94\text{ kJ/mol} & & -285.84\text{ kJ/mol} \\ H^+ & + & OH^- & \longrightarrow & H_2O\,(l) \end{array}$$

$$-285.84 - (-229.94) - (0) = -55.9\text{ kJ/mol}$$

In each of the first four neutralization reactions (Part C), the number of equivalents of acid is in excess of the equivalents of base. The observed temperature change should therefore be about the same in each reaction. The fact that the acid is in excess is emphasized in the fourth reaction by doubling the amount of acid without significantly affecting the amount of temperature change. Although twice as much base is used in the fifth neutralization reaction, it is still the limiting reactant. Therefore, twice as much acid and base are neutralized, resulting in an approximate doubling of the temperature change.

Students need some guidance in interpreting differences between small and large temperature variations. The temperature increases for the first four neutralizations do not give the perfect agreement we might prefer. The differences in the average temperature changes are due in part to small differences in the heat capacities and heats of solution in the acid solutions but are of the same magnitude as the variations between identical runs. Therefore, the argument that the neutralizations in the first four runs are identical type reactions is sound.

Waste requirements: none

EXPERIMENT 26: Distillation of Volatile Liquids

This experiment demonstrates separation techniques, liquids, boiling points, and vapor pressure and provides more experience in graphing. The ethanol used can be any form of denatured ethyl alcohol which does not include methanol. We use the 95% ethanol/5% isopropyl alcohol (IPA) listed but other formulations are available.

We have the equipment setup in advance (Figure 26.1) for students who work in groups of 2–3. Thus, for a class of 24 students, we use 8 setups. Students hook up the water hoses, adjust the hot plates and thermometer, tighten the connections, and add the liquids as instructed in the procedure.

During the prelab session, we emphasize that heating should be slow and gradual. It is difficult to give one set of instructions that will work for every heat source to generate data that fits on a single data table. We have pretested each hot plate or heating mantle to determine the optimum setting for heating the liquids so that it takes at least 3 minutes to go from 34°C to 78°C for the ethanol; and no longer than 12 minutes to go from 34°C to 100°C for heating the water. If this pretesting of the hot plates is not done, it may take longer than the time allotted on the data table to complete each distillation.

Some student guidelines that are useful include: (1) each distillation trial should start at about 34°C which requires cooling the flask and thermometer for the second and third liquids, (2) the hot plates should be set relatively low for the alcohol distillation and may have to be turned up for the water distillation in order to reach a plateau within the time provided on the data tables; (3) the alcohol/water solution should begin with the hot plates on the lower setting and adjustments may be necessary. This is an opportunity for students to go beyond blindly following directions and apply what they have learned.

In the third distillation we use red wine because it is noncarbonated and the pigments are nonvolatile. The clear distillate always surprises some of the students and brings home the value of distillation for the separation of volatile from nonvolatile components in a solution.

A bottle for recycling the leftover ethanol and distillate should be available. This recycled alcohol can be used to prepare the 50/50 mixture.

Waste requirements:
A bottle for recycling ethanol/distilled ethanol should be available.

EXPERIMENT 27: Hydrocarbons

Since highly combustible substances are handled in this experiment, the students should be reminded of the fire hazard. The instructor may wish to demonstrate the combustion of toluene in Part B. This combustion was purposely omitted from student experimentation because copious amounts of black smoke are produced when toluene is burned.

Any similar olefin may be used in place of pentene. Petroleum ether (ligroine, boiling range 60–100°C) may be substituted for heptane.

The instructor is advised to check beforehand the reaction of heptane (or ligroine) with bromine. There is a strong possibility that the reaction will be similar to that of amylene due to alkene contamination. If the alkane available shows alkene contamination, it should not be used in the bromine test.

For Parts C and D, test the kerosene to see that it reacts as a saturated hydrocarbon. We have found that some commercial kerosene does not react as a saturated hydrocarbon.

To obtain the desired results in the combustion of acetylene (Part E), the tube in which the gas is collected must be stoppered or at least removed from over the reacting calcium carbide as soon as the water is displaced from the test tube. Even though there may be very little danger present when burning these small quantities of acetylene, the student should take the suggested precautions.

Waste requirements:
A bottle for Waste Organic Solvents is needed.

EXPERIMENT 28: Alcohols, Esters, Aldehydes, and Ketones

Again the students are handling combustible organic substances in this experiment and should be reminded of the possible fire hazard.

Copper wires (No. 18) with four or five spiral turns at one end are used for the oxidation in Part B.2. The wires should be at least eight inches overall so that they can easily be inserted into and removed from a 150 mm test tube. The wire is dropped into the alcohol at the end of the experiment merely to allow the copper to cool and remain in the reduced state.

The odor of an ester, especially ethyl acetate, may be difficult to detect in reaction mixtures. To overcome this difficulty the amount of acetic acid in the preparation of acetates should be kept to a minimum so that its odor does not overpower that of the ester. The odor of the ester can be detected more readily if a few drops of the reaction mixture are poured onto a piece of filter paper.

A good silver mirror in the Tollens test for aldehydes depends on the test tube being scrupulously clean. Otherwise the silver does not plate out on the test tube walls but appears as a gray to brown precipitate.

Waste requirements:
Two bottles are needed: (1) Waste Organic Solvents and (2) Waste Heavy Metals (MnO_4^-, Ag^+)

D.   REAGENT LISTS

The following lists of reagents and experiments in which they are used will be useful for budgeting chemicals. The numbers given in parentheses following each reagent correspond to the experiments in which it is used. The following solutions, used in the ionization demonstration in Experiment 12, are not included in the list as only about 10 mL of each are needed: 1 M HCl, 1 M $HC_2H_3O_2$, 1 M NaOH, 1 M NH_4OH, and 0.1 M $NaNO_3$, 0.1 M NaBr, 0.1 M $Ni(NO_3)_2$, 0.1 M $CuSO_4$, 0.1 M NH_4Cl.

Detailed directions for the preparation of all required solutions are given in Appendix 2 in the laboratory manual.

SOLIDS

Ammonium chloride, NH_4Cl (8, 24, 25)
Barium chloride, $BaCl_2 \cdot 2\,H_2O$ (7, 8)
Barium sulfate, $BaSO_4$ (8)
Benzoic acid, C_6H_5COOH (6)
Boiling stones,   (20, 26)
Candles   (3)
Calcium carbide (small lumps), CaC_2 (27)
Calcium hydroxide, $Ca(OH)_2$ (12)
Calcium oxide, CaO (12)
Cobalt chloride paper   (7)
Copper strips, Cu (4, 11)
Copper wire, #18, Cu (28); #24   (20)
Copper(II) sulfate pentahydrate, $CuSO_4 \cdot 5\,H_2O$ (7)
Cotton   (3)
Ice   (2, 6)
Iron wire, #20-24, Fe (12)
Lead strips, Pb (11)
Lead(II) iodide, PbI_2 (1)
Magnesium strips, Mg (3, 4, 12)
Magnesium oxide, MgO (12)
Magnesium sulfate, $MgSO_4 \cdot 7\,H_2O$ (7)
Manganese dioxide (powdered), MnO_2 (3, 21)
Marble chips, $CaCO_3$ (12)
pH paper   (4)

Potassium, hydrogen phthalate, $KHC_8H_4O_4$ (22)
Potassium bicarbonate, $KHCO_3$ (7, 15)
Potassium chlorate, C.P., $KClO_3$ (9)
Potassium chloride, C.P., KCl (9)
Sand paper or emery cloth   (11)
Salicylic acid, $C_6H_4(COOH)(OH)$ (28)
Sodium, Na (4)
Sodium bicarbonate, $NaHCO_3$ (7, 12)
Sodium chloride (course crystals), $NaCl$ (8)
Sodium chloride (fine crystals), $NaCl$ (1, 2, 8, 12)
Sodium peroxide, Na_2O_2 (3)
Sodium sulfate, Na_2SO_4 (8)
Sodium sulfite, Na_2SO_3 (10)
Steel wool, Fe(Grade 0 or 1)   (3, 4)
Strontium chloride, $SrCl_2 \cdot 6\,H_2O$ (7) (unknown)
Styroform cups, 6 oz.   (5, 25)
Sucrose, $C_{12}H_{22}O_{11}$ (12)
Sulfur, S (3, 12)
Wood splints   (3, 4, 12, 27)
Zinc, mossy, Zn (4)
Zinc strips, Zn (0.01 inch thick),   (4, 11)
Zinc sulfate, $ZnSO_4 \cdot 7\,H_2O$ (7)

PURE LIQUIDS/COMMERCIAL MIXTURES

Acetic acid (glacial), CH_3COOH (6, 28)
Acetone, CH_3COCH_3 (20, 28)
Decane, $C_{10}H_{22}$ (8, 13)
Ethanol 95%/5% Isopropyl alcohol   (26)
Ethyl alcohol (ethanol), 95%, C_2H_5OH (20, 28)
Glycerol   (1)
Heptane (or low boiling petroleum ether), C_7H_{16} (27)

Isoamyl alcohol (3-methyl-1-butanol), $C_5H_{11}OH$ (28)
Isopropyl alcohol (2-propanol), C_3H_7OH (8, 28)
Kerosene (alkene free)   (8, 27)
Methyl alcohol (methanol), CH_3OH (20, 28)
Pentene (amylene), C_5H_{10} (27)
Toluene, $C_6H_5CH_3$ (27)
1,1,1-Trichloroethane, CCl_3CH_3 (27)

SOLUTIONS

Acetic acid, concentrated (glacial), $HC_2H_3O_2$ (6, 28)
Acetic acid, dilute, 6 M, $HC_2H_3O_2$ (4, 12)
Ammonium chloride, 0.1 M, NH_4Cl (10)

Ammonium chloride, saturated, NH_4Cl (24)
Ammonium hydroxide, concentrated, NH_4OH (12, 14)

Ammonium hydroxide, dilute, 6 M,
 NH$_4$OH   (10, 24, 28)
Barium chloride, 0.10 M,
 BaCl$_2 \cdot$ 2 H$_2$O   (10, 13)
Barium hydroxide, saturated,
 Ba(OH)$_2 \cdot$ 8 H$_2$O   (12)
Bromine in 1,1,1-trichloroethane, 5%
 solution, Br$_2$ (27)

Calcium chloride, 0.1 M, CaCl$_2 \cdot$ 2 H$_2$O   (10)
Chlorine water, Cl$_2$ (13)
Cobalt(II) chloride, 0.1 M,
 CoCl$_2 \cdot$ 6 H$_2$O   (24)
Copper(II) nitrate,   0.1 M,
 Cu(NO$_3$)$_2 \cdot$ 3 H$_2$O   (11)
Copper(II) sulfate, 0.1 M,
 CuSO$_4 \cdot$ 5 H$_2$O   (4, 10, 12, 24)

Formaldehyde, 10% solution,
 H$_2$CO   (28)

Glucose, 10% solution, C$_6$H$_{12}$O$_6$ (28)

Hydrochloric acid, concentrated,
 12 M HCl   (10, 24)
Hydrochloric acid, dilute, 3 M, HCl   (25)
Hydrochloric acid, dilute, 6 M,
 HCl   (4, 10, 12, 13, 14)
Hydrochloric acid, dilute, 0.1 M,
 0.01 M, 0.001 M   (12)
Hydrogen peroxide, 3%, H$_2$O$_2$ (21)
Hydrogen peroxide, 9%, H$_2$O$_2$ (3)
Hydrogen peroxide, 30% H$_2$O$_2$ (for dilution)
Iodine water, saturated, I$_2$ (8, 24)
Iron(III) chloride, 0.1 M,
 FeCl$_3 \cdot$ 6 H$_2$O   (10, 24)

Lead(II) nitrate, 0.1 M, Pb(NO$_3$)$_2$ (1, 11, 14)

Magnesium sulfate, 0.1 M,
 MgSO$_4 \cdot$ 7 H$_2$O   (11)
Mercury(I) nitrate, 0.1 M,
 HgNO$_3 \cdot$ H$_2$O   (14)

Nickel (II) nitrate, 0.1 M,
 Ni(NO$_3$)$_2 \cdot$ 6 H$_2$O   (12, 16)
Nitric acid, dilute, 3 M HNO$_3$ (25)
Nitric acid, dilute, 6 M,
 HNO$_3$ (9, 10, 12, 13, 14, 24)

Phenolphtalein, 0.2% solution,
 in ethanol-water   (4, 12, 22, 23, 24)
Phosphoric acid, dilute, 3 M, H$_3$PO$_4$ (4)
Potassium chloride, saturated, KCl   (8)
Potassium nitrate, 0.1 M, KNO$_3$ (10)
Potassium permanganate, 0.1 M,
 KMnO$_4$ (27, 28)
Potassium permanganate, 0.002 M,
 KMnO$_4$ (16)
Potassium thiocyanate, 0.1 M, KSCN   (24)

Silver nitrate, 0.10 M, AgNO$_3$
 (9, 10, 11, 13, 14, 24, 28)
Sodium bromide, 0.1 M, NaBr   (12, 13)
Sodium carbonate, 0.1 M, Na$_2$CO$_3$ (10, 13)
Sodium chloride, 0.1 M, NaCl   (10, 13)
Sodium chloride, saturated solution, NaCl   (24)
Sodium hydroxide, 10% solution,
 NaOH   (10, 12, 24, 28)
Sodium hydroxide, 1.25 M, NaOH   (25)
Sodium hydroxide for unknowns,
 NaOH   (22, 23)
Sodium iodide, 0.1 M, NaI   (1, 13, 14)
Sodium phosphate, 0.1 M, Na$_3$PO$_4$ (13)
Sodium sulfate, 0.1 M, Na$_2$SO$_4$ (13)
Sulfuric acid, concentrated, 18 M,
 H$_2$SO$_4$ (25, 28)
Sulfuric acid, 9 M, H$_2$SO$_4$ (4)
Sulfuric acid, dilute, 3 M H$_2$SO$_4$
 (4, 10, 11, 12, 24, 28)

Vinegar, commercial (colorless), HC$_2$H$_3$O$_2$ (23)

Wine, red   (26)

Zinc nitrate, 0.1 M, Zn(NO$_3$)$_2$ (10)

The following list of reagents shows the amount needed for 100 students in each experiment. A 50% excess over the actual amount specified in the experiment is included to allow for wastage.

	Amount per 100 Students		Amount per 100 Students
Experiment 1		**Experiment 1 (continued)**	
PbI$_2$, specimen sample	10 g	Pb(NO$_3$)$_2$, 0.1 M	1000 mL (33 g)
NaNO$_3$, specimen sample	10 g	NaI, 0.1 M	500 mL (1.5 g)
NaCl	300 g	Glass rod	100 ft
Glycerol	25 mL	Glass tubing, 6mm	400 ft

Experiment 2

Ice	—
NaCl, coarse	800 g
NaCl, fine	300 g
Rulers	100
Solid objects for density (see Section C)	—

Experiment 3

Büchner funnel-vacuum flask setup	1 or 2
Candles	50
Mg strips	28 g
MnO_2, powdered	300 g
H_2O_2, 9%	7.5 L (2.25 L of 30%)
Steel wool, fine	150 g
Sulfur, roll	75 g
Wood splints	600
Demonstration	
Cotton	—
Na_2O_2	—

Experiment 4

Cu strips	75 g
Mg strips	28 g
Zn strips	25 g
Na metal	20 g
Steel wool	150 g
Zn, mossy	1500 g
$HC_2H_3O_2$, 6 M	300 mL (105 mL conc.)
$CuSO_4$, 0.1 M	300 mL (7.5 g)
HCl, 6 M	3 L (1.5 L conc.)
H_3PO_4, 3 M	600 mL (120 mL 85%)
H_2SO_4, 3 M,	600 mL (100 mL conc.)
H_2SO_4, 9 M	2000 mL
pH paper	4 packages of strips
Phenolphthalein, 0.2% solution	200 mL
Wood splints	600

Experiment 5

Styrofoam cups, 6 oz	150
Metal samples to fit into a large test tube (Al, Cu, Fe, Pb)	—
4 × 4 in. cardboard squares with thermometer hole	50

Experiment 6

$HC_2H_3O_2$, glacial	1500 mL
Benzoic acid, C_6H_5COOH	75 g
Crushed ice	—

Experiment 7

$CuSO_4 \cdot 5 H_2O$	600 g

Experiment 7 (continued)

$CoCl_2$ test paper	Two 8 × 11 sheets
Assorted unknowns	700 g (See Sect. C Notes)

Experiment 8

NH_4Cl	360 g
$BaCl_2$	15 g
$BaSO_4$	15 g
Na_2SO_4	15 g
NaCl, course	300 g
NaCl, fine	800 g
$C_{10}H_{22}$, decane	300 mL
Isopropyl alcohol	300 mL
Kerosene	300 mL
$I_2 - H_2O$	750 mL (5 g I_2)
KCl (saturated)	900 mL (850 g)

Experiment 9

$KClO_3$, C. P.	750 g
KCl, C. P.	15 g
HNO_3, 6 M	250 mL (100 mL conc.)
$AgNO_3$, 0.1 M	450 mL (8 g)

Experiment 10

NH_4OH, 6 M	750 mL (300 mL conc.)
NH_4Cl, 0.1 M	450 mL (2.5 g)
$BaCl_2$, 0.1 M	450 mL (11 g)
$CaCl_2$, 0.1 M	450 mL (6.6 g)
$CuSO_4$, 0.1 M	900 mL (22.5 g)
HCl, 6 M	450 mL (225 mL conc.)
HCl, 12 M, conc.	600 mL conc.
$FeCl_3$, 0.1 M	450 mL (12 g)
HNO_3, 6 M	450 mL (170 mL conc.)
KNO_3, 0.1 M	450 mL (4.5 g)
$AgNO_3$. 0.1 M	450 mL (7.6 g)
Na_2CO_3, 0.1 M	900 mL (10 g)
NaCl, 0.1 M	900 mL (5.5 g)
NaOH, 10%	900 mL (100 g)
Na_2SO_3	150 g
H_2SO_4, 3 M	900 mL (150 mL conc.)
$Zn(NO_3)_2$, 0.1 M	45 mL (14 g)

Experiment 11

Cu strips	75 g
Pb strips	75 g
Zn strips	100 g
$Cu(NO_3)_2$, 0.1 M	600 mL (18 g)
$Pb(NO_3)_2$, 0.1 M	600 mL (25 g)
$MgSO_4$, 0.1 M	600 mL (19 g)
$AgNO_3$, 0.1 M	600 mL (13 g)
H_2SO_4, 3 M	1.5 L (250 mL conc.)
Sandpaper	—

Experiment 12

Demonstration	*See Instructor's Manual,* Exp. 12
CaO	15 g
$Ca(OH)_2$	15 g
Fe wire	—
Mg strips	30 g
MgO	15 g
$CaCO_3$ (marble chips)	525 g
$NaHCO_3$	750 g
S	15g
$HC_2H_3O_2$, 6 M	775 mL (275 mL conc.)
HCl, 6 M	3 L (1.5 L conc.)
HCl, 0.10 M	300 mL
HCl, 0.010 M	300 mL
HCl, 0.0010 M	300 mL
HNO_3, 6 M	750 mL (280 mL conc.)
NH_4OH, conc.	75 mL
H_2SO_4, 3 M	750 mL (125 mL conc.)
NaOH, 10%	90 mL (11 g)
Wood splints	400
Phenolphthalein, 0.2% solution	50 mL

Experiment 13

$C_{10}H_{22}$, Decane	2700 mL
$BaCl_2$, 0.1 M	2700 mL (66 g)
Cl_2 water	2700 mL
HCl, 6 M	4 L (2 L conc.)
HNO_3, 6 M	4 L (1.5 L conc.)
$AgNO_3$, 0.1 M	1300 mL (22 g)
NaBr, 0.1 M	900 mL (5.3 g)
Na_2CO_3, 0.1 M	900 mL (9.5 g)
NaCl, 0.1 M	900 mL (5.3 g)
NaI, 0.1 M	900 mL (13.5 g)
Na_3PO_4, 0.1 M	900 mL (34 g)
Na_2SO_4, 0.1 M	900 mL (12.8 g)
Unknown solutions	See Sect. C Notes

Experiment 14

NH_4OH, conc.	600 mL
HCl, 6 M	600 mL (300 mL conc.)
$Pb(NO_3)_2$, 0.1 M	300 mL (9.9 g)
$Hg_2(NO_3)_2$, 0.1 M	300 mL (8.4 g)
HNO_3, 6 M	1000 mL (375 mL conc.)
NaI	200 mL (5 g)
$AgNO_3$, 0.1 M	300 mL (5.1 g)
"Known" solution	300 mL
Unknown solutions	—

Experiment 15

$KHCO_3$	450 g
HCl, 6 M	900 mL

Experiment 16

$Ni(NO_3)_2$, 0.1 M	200 mL (6 g)
$KMnO_4$, 0.002 M	200 mL (0.06 g)
Hand-held spectroscopes	1 per student
1.7 meter wave motion spring (CENCO 384740G)	1 per 5 students
Spectrophotometer with cuvettes	3 per lab
Vapor lamps (H_2 and Ne) with power supplies	2 each per lab
Incandescent (60–100 watt) lights	2 per lab
Fluorescent lights (or ceiling lights)	2 per lab
Spectrum chart	1 per lab
Colored pencils (red, orange, yellow, blue, violet)	6 sets per lab
Meter stick	5 per lab
Stopwatch (recommended)	1 per wave motion spring

Experiment 17

Molecular Model Sets	1 per 2 students
Sources:	
Carolina (Prentice-Hall): D8-84-0160 or D8-84-0164	
Fisher: CLS44052	
VWS: WLS-61815	
Many others have equivalent sets	

Experiment 18

Elasticity of Gases Kit	1 kit per 2–3 students
or Simple Boyle's Law Apparatus	
Sources:	
Cenco: 72710-25G	
Fisher: CLS52009	
VWR: WL1077	
Silicone grease	1 tube
0.5 kg and 1 kg slotted masses	12 per lab
Bricks (assorted masses)	18 per lab
Vernier calipers	1 per 2–3 students
Balance, large capacity (to 2.5 kg)	1 per lab

Experiment 19

No special equipment or reagents needed

Experiment 20

Acetone	75 mL
Ethanol	75 mL
Methanol	75 mL
Cu wire, #28	50 ft
Boiling stones	1 bottle
One gallon can per class	—

	Amount per 100 Students			Amount per 100 Students

Experiment 21

H_2O_2, 3.0% solution	1500 mL
MnO_2, powdered	200 g
Disposable syringe (reusable)	75
Needles for syringe (reusable)	75
Rubber stopper (new and soft)	75
2-liter beakers or battery jars	15
Büchner funnel-vacuum flask setup	1 or 2 per class

Experiment 22

$KHC_8H_4O_4$ (KHP)	350 g
Phenolphthalein, 0.2% solution	50 mL
NaOH unknowns	100–250 mL/student
(See Section C, Exp. 22 Notes)	

Experiment 23

Acid unknowns	50 mL/student
(See Section C, Exp. 23 Notes)	
NaOH (from Exp. 22)	—
Phenolphthalein, 0.2% solution	100 mL
Vinegar	6000 mL

Experiment 24

NH_4Cl	300 g
NH_4Cl, saturated	450 mL (27 g)
NH_4OH, 6 M	200 mL
$CoCl_2$, 0.1 M	900 mL (21.4 g)
$CuSO_4$, 0.1 M	400 mL (15 g)
$FeCl_3$, 0.1 M	750 mL (20.3 g)
HCl, conc.	600 mL
Phenolphthalein, 0.2% solution	50 mL
KSCN, 0.1 M	750 mL (7.3 g)
$AgNO_3$, 0.1 M	150 mL (2.6 g)
NaCl, saturated	450 mL (27 g)
H_2SO_4, 3 M	100 mL

Experiment 25

NH_4Cl	450 g
HCl, 3 M	6 L
HNO_3, 3 M	1.5 L
NaOH, 1.25 M	9 L (450 g)
H_2SO_4, conc.	750 mL
Styrofoam cups, 6 oz.	200

Experiment 26

Boiling stones	1 bottle
Distillation setup	1 setup per 3 students
125 mL or 250 mL flask (Fig. 26.1)	
Hot plate or heating mantle	1 per each setup
Denatured alcohol (95% ethanol–5% isopropyl alcohol) (denaturant must not be methanol)	2.5 L
Ethanol-water (50–50 solution)	2.5 L
Red wine (uncarbonated)	2.5 L
pot holders or mitts	1 for each setup

Experiment 27

Calcium carbide (small lumps), CaC_2	300 g
Pentene (amylene), C_5H_{10}	750 mL
Heptane (or pet. ether), C_7H_{16}	750 mL
Kerosene (alkene free)	450 mL
Br_2, 5% in CCl_3CH_3	100 mL
$KMnO_4$, 0.1 M	200 mL (3.2 g)
Toluene, $C_6H_5CH_3$	750 mL
Wood splints	600

Experiment 28

Cu wire, #18 with spiral	360 cm
Salicylic acid	75 g
$HC_2H_3O_2$, glacial	300 mL
Acetone	15 mL
Ethyl alcohol (95%)	450 mL
Isoamyl alcohol (3-methyl-1-butanol)	350 mL
Isopropyl alcohol	600 mL
Methyl alcohol	1000 mL
NH_4OH, 6 M	150 mL (60 mL conc.)
Glucose, 10%	35 mL (3.5 g)
Formaldehyde, 10%	15 mL
$KMnO_4$, 0.1 M	45 mL (1 g)
$AgNO_3$, 0.1 M	1.2 L (20.4 g)
NaOH, 10%	45 mL (5.0 g)
H_2SO_4, 3 M	45 mL (7.5 mL conc.)
H_2SO_4, conc.	200 mL

NAME **KEY**

SECTION _____ DATE _____

REPORT FOR EXPERIMENT 1

INSTRUCTOR _____

Laboratory Techniques

A, B. Laboratory Burners and Glassworking

Glassware shown in Figure 1.3

Articles	Instructor's Check and Comments
Straight tubes (2)	
Right-angle bends (2)	
Delivery tube (1)	
Buret tips (2)	(optional)
Stirring rod (1)	

Instructor's OK or grade on glass work _____

QUESTIONS AND PROBLEMS

1. Why is it necessary to turn off the gas with the gas cock rather than with the valve on the burner?

   To prevent gas leaks (due to cracked or ruptured tubing).

2. Why is air mixed with gas in the barrel of the burner before the gas is burned?

   To give better combustion or to avoid smoky flames or to give a nonluminous flame or to control the temperature of the flame.

3. How would you adjust a burner which

   (a) has a yellow and smoky flame?

   Increase air flow by opening air vents.

   (b) is noisy with a tendency to blow itself out?

   Decrease air flow by closing air vents.

– 9 –

4. Why are glass tubes and rods always fire-polished after cutting?

   To prevent personal injury <u>or</u> to remove rough edges (and to prevent gouging of rubber stoppers).

5. Explain briefly how to insert glass tubing into a rubber stopper.

   Apply a drop(s) of lubricant to the stopper hole, grip the fire-polished glass tubing close to the stopper, insert and twist gently. Do not force the glass into the stopper.

6. Name the lubricant used for inserting glass tubing in rubber stoppers.

   _____ **glycerol** _____

C. Evaporation

Give the name and formula of the residue remaining after evaporation:

Name _____ **Sodium Chloride** _____ Formula _____ **NaCl** _____

D. Filtration

1. What is the name, formula, and color of the precipitate recovered by filtration?

   Name ____ **Lead (II) iodide** ____ Formula ____ **PbI₂** ____ Color ____ **Yellow** ____

2. Explain why the filter paper with the precipitate is collected in a jar instead of thrown into the trash can? (Refer to the section on waste disposal in Laboratory Rules and Safety Practices.)

   The precipitate is lead (II) iodide. It is considered toxic and should not be disposed of with nontoxic substances where it could contaminate the environment.

3. Give the names and formulas of two compounds that must be present in the filtrate.

   Name _____ **Sodium nitrate** _____ Formula _____ **NaNO₃** _____

   Name _____ **Water** _____ Formula _____ **H₂O** _____

REPORT FOR EXPERIMENT 2

Measurements

> ✓ Student results will be variable and will not exactly match the typical student data shown on the key as a guideline for grading.

A. Temperature

1. Water at room temperature | **26.3°C** °C
2. Boiling point | **101.1°C** °C
3. Ice water
   Before stirring | **10.5°C** °C
   After stirring for 1 minute | **4.0°C** °C
4. Ice water with salt added | **2.0°C** °C

B. Mass

> ✓ Sample mass data measured on a balance with precision to 0.0001 g.

1. 250 mL beaker | **101.2788** g
2. 125 mL Erlenmeyer flask | **95.2508** g
3. Weighing paper or weighing boat | **(0.4048) or 2.0864** g
4. Mass of weighing paper/boat + sodium chloride | **4.1457** g

Mass of sodium chloride (show calculation setup) | **2.0593** g

$$4.1457 \text{ g}$$
$$-2.0864 \text{ g}$$
$$\overline{2.0593 \text{ g}}$$

C. Length

1. Length of ⟵————————⟶ | $2\frac{3}{16}$ in. | **5.56** cm
2. Height of 250 mL beaker | $3\frac{7}{16}$ in. | **8.73** cm
3. Length of test tube | $6\frac{1}{8}$ in. | **15.56** cm

D. Volume

1. Test tube | **28.6** mL
2. 125 mL Erlenmeyer flask | **144** mL
3. Height of 5.0 mL of water in test tube | **3.4** cm
4. Height of 10.0 mL of water in test tube | **5.8** cm

E. Density

1. Density of Water

Mass of empty graduated cylinder   67.5392 g

Volume of water   50.0 mL

Mass of graduated cylinder and water   115.7126 g

Mass of water (show calculation setup)   115.7126 g   48.1734 g
                                        − 67.5392 g
                                          481734 g

Density of water (show calculation setup) $\dfrac{48.1734 \text{ g}}{50.0 \text{ mL}} = 0.963 \dfrac{\text{g}}{\text{mL}}$ $0.963 \dfrac{\text{g}}{\text{mL}}$

                                                                 (range 0.900 − 1.100)

2. Density of a Rubber Stopper

Mass of rubber stopper   3.9894 g

Initial volume of water in cylinder   25.0 mL

Final volume of water in cylinder   28.1 mL

Volume of rubber stopper (show calculation setup)   28.1 mL   3.1 mL
                                                  −25.0 mL
                                                    3.1 mL

Density of rubber stopper (show calculation setup) $\dfrac{3.9894 \text{ g}}{3.1 \text{ mL}} = 1.3 \dfrac{\text{g}}{\text{mL}}$ $1.3 \dfrac{\text{g}}{\text{mL}}$

                                                                 (range 1.1 − 1.5)

3. Density of a Solid Object

Number of solid object   (Aluminum)

Mass of solid object   18.2909 g

Initial volume of water in graduated cylinder   25.6 mL

Final volume in graduated cylinder   31.3 mL

Volume of solid object (show calculation setup)   31.3 mL   5.7 mL
                                                 −25.6 mL
                                                   5.7 mL

Density of solid object (show calculation setup) $\dfrac{18.2909 \text{ g}}{5.7 \text{ mL}} = 3.2 \dfrac{\text{g}}{\text{mL}}$ $3.2 \dfrac{\text{g}}{\text{mL}}$

QUESTIONS AND PROBLEMS

1. The directions state "weigh about 5 grams of sodium chloride". Give minimum and maximum amounts of sodium chloride that would satisfy these instructions.

   4.8 g to 5.2 g

2. Two students each measured the density of a quartz sample three times:

	Student A	Student B	
1.	3.20 g/mL	2.82 g/mL	The density found in the *Handbook*
2.	2.58 g/mL	2.48 g/mL	*of Chemistry and Physics* for quartz
3.	2.10 g/mL	2.10 g/mL	is 2.65 g/mL
mean	2.63 g/mL	2.63 g/mL	

   (a) Which student measured density with the greatest precision? Explain your answer.

   Student B because the density values for each trial are within 0.2 of the mean. For Student A the density measurements were ±0.5 for trials 1 and 3.

   (b) Which student measured density with the greatest accuracy? Explain your answer.

   Both students had the same accuracy because their mean density measurements are close to the accepted value of 2.65 g/mL.

Show calculation setups and answers for the following problems.

3. Convert 21°C to degrees Fahrenheit.   _____**70.°F**_____

   $$°F = 1.8(21°C) + 32 = 38 + 32 = 70.°F$$

4. Convert 101°F to degrees Celsius.   _____**38°C**_____

   $$°C = \frac{101°F - 32}{1.8} = \frac{69}{1.8} = 38°C$$

5. An object is 9.6 cm long. What is the length in inches?   _____**3.8 in.**_____

   $$9.6 \, cm\left(\frac{1 \, in.}{2.54 \, cm}\right) = 3.8 \, in.$$

6. An empty graduated cylinder weighs 82.450 g. When filled to 50.0 mL with an unknown liquid it weighs 110.810 g. What is the density of the unknown liquid?

   _____$0.567\frac{g}{mL}$_____

   110.810 g − 82.450 g = 28.360 g (mass of liquid)

   $$\frac{28.360 \, g}{50.0 \, mL} - 0.567\frac{g}{mL}$$

7. It is valuable to know that 1 milliliter (mL) equals 1 cubic centimeter (cm³ or cc). How many cubic centimeters are in an 8.00 oz bottle of cough medicine? (1.00 oz = 29.6 mL)

$$\text{cm}^3 = (8.00 \, \text{oz})\left(\frac{29.6 \, \text{mL}}{1 \, \text{oz}}\right)\left(\frac{1 \, \text{cm}^3}{1 \, \text{mL}}\right)$$

_____ 237 cm³

8. A metal sample weighs 56.8 g. How many ounces does this sample weigh? (1 lb = 16 oz)

$$\text{oz} = (56.8 \, \text{g})\left(\frac{1 \, \text{lb}}{454 \, \text{g}}\right)\left(\frac{16 \, \text{oz}}{1 \, \text{lb}}\right) = 2.00 \, \text{oz}$$

_____ 2.00 oz

9. Convert 15 nm into km.

$$\text{km} = (15 \, \text{nm})\left(\frac{10^{-9} \, \text{m}}{1 \, \text{nm}}\right)\left(\frac{1 \, \text{km}}{10^3 \, \text{m}}\right)$$

_____ 1.5×10^{-11} km

REPORT FOR EXPERIMENT 3

Preparation and Properties of Oxygen

A. and B.   Generation and Collection of Oxygen

1. What evidence did you observe that oxygen is not very soluble in water?

   Oxygen gas displaced the water <u>or</u> bubbled through and did not dissolve in the water.

2. What is the source of oxygen in the procedure you used?

   Name _____ **Hydrogen peroxide** _____   Formula _____ **H_2O_2** _____

3. What purpose does the manganese dioxide serve in this preparation of oxygen?

   As a catalyst (to speed up the reaction).

4. What gas was in the apparatus before you started generating oxygen? Where did it go?

   Air. As oxygen was generated, this air was swept into the oxygen collecting bottles.

5. What is different about the composition of the first bottle of gas collected compared to the other four?

   The first bottle contains more air than the others. (Most of the air originally present in the generator and delivery tube.)

6. Why are the bottles of oxygen stored with the mouth up?

   To decrease the chance of oxygen escaping from them <u>or</u> because oxygen is slightly denser than air.

7. (a) What is the symbol of the element oxygen?   _____ **O** _____

   (b) What is the formula for oxygen gas?   _____ **O_2** _____

8. Which of the following formulas represent oxides? (Circle) (MgO,) $KClO_3$, (SO$_2$,) (MnO$_2$,) O_2, NaOH, (PbO$_2$,) (Na$_2$O$_2$)

9. Write the word and formula equations for the preparation of oxygen from hydrogen peroxide.

   Word Equation: **Hydrogen peroxide $\longrightarrow$ Water + Oxygen**

   Formula Equation: **$2\,H_2O_2 \longrightarrow 2\,H_2O + O_2$**

10. What substances, other than oxygen, are in the generator when the decomposition of H_2O_2 is complete?

    H_2O and MnO_2

C. Properties of Oxygen

1. Write word equations for the chemical reactions that occurred. (See Study Aid 2.)

   C.1. Combustion of wood. Assume carbon is the combustible material.

   Carbon + Oxygen $\longrightarrow$ Carbon dioxide

   C.2. Combustion of sulfur.

   Sulfur + Oxygen $\longrightarrow$ Sulfur dioxide

   C.5. Combustion of steel wool (iron). (Call the product iron oxide.)

   Iron + Oxygen $\longrightarrow$ Iron oxide

   C.6. Combustion of magnesium.

   Magnesium + Oxygen $\longrightarrow$ Magnesium oxide

2. Write formula equations for these four chemical reactions.

   C.1. (CO_2 is the formula for the oxide of carbon that is formed.)

   $C + O_2 \longrightarrow CO_2$

   C.2. (SO_2 is the formula for the oxide of sulfur that is formed.)

   $S + O_2 \longrightarrow SO_2$

C.5. (Fe$_3$O$_4$ is the formula for the oxide of iron that is formed.)

3 Fe + 2 O$_2$ $\longrightarrow$ Fe$_3$O$_4$

C.6. (MgO is the formula for the oxide of magnesium that is formed.)

2 Mg + O$_2$ $\longrightarrow$ 2 MgO

3. Combustion of a candle.

   (a) Number of seconds that the candle burned in the bottle of oxygen.   __Variable__

   3(a) should be at least three times 3(b).

   (b) Number of seconds that the candle burned in the bottle of air.   __Variable__

   (c) Explain this difference in combustion time.

   The candle burned longer in the bottle of oxygen because this bottle contained a higher concentration of oxygen. or It took longer to use up the oxygen in 3(a) because there was more available.

   (d) Is it scientifically sound to conclude that all the oxygen in the bottle was reacted when the candle stopped burning? Explain.

   No. Further testing was not done to determine if all the oxygen was consumed.

4. What were the results of the experiment in which a bottle of oxygen was placed over a bottle of air? Explain the results.

   The splint burst into flame in both bottles; therefore, they both contained more oxygen than normal air. The two gases, oxygen and air, mixed (diffused) during the time they were allowed to stand mouth to mouth.

5. (a) Describe the material that is formed when magnesium is burned in air.

   White, powdery solid or grayish white powdery solid (white smoke)

   (b) What elements are in this product?

   Magnesium and oxygen

6. (a) What is your conclusion about the rate or speed of a chemical reaction with respect to the concentration of the reactants—for example, a combustion in a high concentration of oxygen (pure oxygen) compared to a combustion in a low concentration of oxygen (air)?

Reactions proceed faster in a higher concentration of reactants. <u>or</u> Reactions proceed faster in a higher concentration of oxygen.

(b) What evidence did you observe in the burning of sulfur to confirm your conclusion in 6(a)?

The sulfur burned with a brighter (more intense) blue flame in oxygen than in air. <u>or</u> The sulfur burned faster in oxygen than in air.

REPORT FOR EXPERIMENT 4

Preparation and Properties of Hydrogen

A. Preparing Hydrogen from Water

1. Describe what you observed when sodium was dropped into water.

   Sodium flitted around on the surface and slowly disappeared (with fizzing) or a reasonably similar answer.

2. Describe what you observed when a flame was brought to the mouth of the test tube.

   A popping or "barking" sound (also a flash) or a small explosion (and water vapor condensed on the inside of the test tube).

3. (a) What color did the litmus papers turn?   ____blue____

   (b) What color did the phenolphthalein turn?   ____pink____

   (c) What was the pH of the solution?   ____13____

4. Did the reaction make the solution acidic or basic?   ____basic____

5. Complete and balance the following word and formula equations:

   Sodium + Water $\longrightarrow$ **Sodium hydroxide + Hydrogen**

   $2\,Na + 2\,H_2O \longrightarrow 2\,NaOH + H_2$

B. Preparing Hydrogen from Acids

1. (a) Write the symbols of the metals that reacted with dilute hydrochloric acid.

   Zn, Fe, Mg

   (b) Write the symbols of the four metals in order of decreasing ability to liberate hydrogen from hydrochloric acid.

   ____**Mg**____   ____**Zn**____   ____**Fe**____   ____**Cu**____

2. Write the formulas of the four acids in order of decreasing strength based on the rates at which they displace hydrogen when reacting with zinc. Record the pH below each formula.

formula	HCl	H_2SO_4	H_3PO_4	$HC_2H_3O_2$
pH	0	0	1	2

C. Generation and Collection of Hydrogen

1. Why was water added to cover the bottom of the thistle tube?

   To prevent the escape of hydrogen through the thistle tube.

2. What do we call this method of collecting a gas?

   Downward displacement of water <u>or</u> displacement of water.

3. What physical property of hydrogen, other than that it is less dense than water, allows it to be collected in this manner?

   Hydrogen is insoluble in water <u>or</u> is slightly soluble in water.

D. Properties of Hydrogen

1. What happened when the splint was brought to the mouth of the first bottle of gas collected?

   An explosion occurred, accompanied by a loud popping or "barking" sound. (A flash also occurred.)

2. Describe fully the results of testing the second bottle of gas.

   There was an initial soft "pop." The splint was extinguished when thrust into the bottle, and reignited when partially withdrawn.

3. (a) Is hydrogen combustible?   _____ **Yes** _____

   (b) What evidence do you have from testing the second bottle of gas?

   The initial pop <u>or</u> the hydrogen burning at the mouth of the bottle caused the splint to reignite.

4. (a) Does hydrogen support combustion?   _____ **No** _____

   (b) What evidence do you have of this from testing the second bottle of gas?

   The burning splint was extinguished when it was thrust into the bottle of pure hydrogen.

5. What compound was formed during the testing of the first and second bottles of gas?

_____**water**_____

6. Why did the first bottle of gas behave differently from the second bottle?

The first bottle contained a mixture of air and hydrogen. The second bottle, only hydrogen.

7. (a) What happened when the splint was brought to the mouth of the third bottle of gas?

Nothing (no evidence of hydrogen).

(b) How do you account for this?

With the bottle left open hydrogen had escaped <u>or</u> hydrogen is less dense than air.

8. When testing the fourth bottle:

(a) What was the result with the top bottle?

Explosion <u>or</u> popping sound.

(b) What was the result with the lower bottle?

Explosion <u>or</u> popping sound.

(c) How do you account for these results?

The air and hydrogen had mixed (diffused).

9. Complete the following word equations:

Zinc + Sulfuric acid $\longrightarrow$ **Zinc sulfate + Hydrogen**

Magnesium + Hydrochloric acid $\longrightarrow$ **Magnesium chloride + Hydrogen**

Hydrogen + Oxygen $\longrightarrow$ **Water**

10. Complete and balance the following corresponding formula equations:

$Zn(s) +$ $H_2SO_4(aq)$ $\longrightarrow$ **$ZnSO_4 + H_2(g)$**

$Mg(s) + 2\,HCl(aq)$ $\longrightarrow$ **$MgCl_2 + H_2(g)$**

$2\,H_2(g) +$ $O_2(g)$ $\longrightarrow$ **$2\,H_2O$**

QUESTIONS AND PROBLEMS

1. What does pH measure?

   H^+ ion concentration $\left(\text{in } \dfrac{\text{moL}}{\text{L}} \right)$

2. What gave better evidence of acid strength, evolution of hydrogen or pH?

   Evolution of hydrogen.

NAME _____**KEY**_____

SECTION _____ DATE _____

INSTRUCTOR _____

REPORT FOR EXPERIMENT 5

Calorimetry and Specific Heat

> ✓ Student results will be variable and will not exactly match the typical student data shown on the key as a guideline for grading.

Measurements and Calculations

	TRIAL 1	TRIAL 2
1. Mass of metal sample	17.4697 g	17.4697 g
2. Mass of calorimeter (styrofoam cups)	7.0390 g	7.0329 g
3. Mass of styrofoam cups + water	70.5280 g	72.9330 g
4. Mass of water (show calculation setup)	63.4890 g	65.9001 g
5. Initial water temperature	22.0°C	11.0°C
6. Temperature of heated metal sample	98.2°C	99.9°C
7. Final temperature of water and metal	26.5°C	17.3°C
8. Change in temperature, Δt_w of the water in the calorimeter (show calculation setup)	4.5°C	6.3°C
9. Change in temperature, Δt_x of the metal sample (show calculation setup)	71.7°C	82.6°C
10. Specific heat ($sp\ ht_w$) of water	4.184 J/g°C	4.184 J/g°C
11. Heat (q_w) gained by water (show calculation setup)	1195 J	1737 J
12. Heat (q_x) lost by metal sample	1195 J	1737 J
13. Specific heat ($sp\ ht_x$) of the metal (experimental value) (show calculation setup)	0.954 J/g°C	1.20 J/g°C

4. (Mass of water calculation setup)

$$\begin{array}{r} 70.5280 \\ -7.0390 \\ \hline 63.4890 \end{array}$$

8. $26.5°C - 22.0°C = 4.5°C$

9. $98.2°C - 26.5°C = 71.7°C$

11. $(63.4890\ \text{g})\left(\dfrac{4.184\ \text{J}}{\text{g}°\text{C}}\right)(4.5°\text{C}) =$

13. $\dfrac{1195\ \text{J}}{(17.4697\ \text{g})(71.7°\text{C})} =$

14. Name of metal __**Aluminum**__   Theoretical Specific Heat __0.900 J/g°C__

QUESTIONS AND PROBLEMS

1. Why is it important for there to be enough water in the calorimeter to completely cover the metal sample?

   So all the heat from the metal will be absorbed by the water and none transferred to the air.

2. Why did we heat the metal in a dry test tube rather than in the boiling water?

   So any hot water clinging to the metal will not mix with and add heat to the water in the calorimeter.

3. The water in the beaker gets its heat energy from the ____**burner**____ and the water in the calorimeter gets its heat energy from the __**hot metal sample**__.

4. What is the specific heat in J/g°C for a metal sample with a mass of 95.6 g which absorbs 841 J of energy when its temperature increases from 30.0°C to 98.0°C?

   $$\text{sp. ht.} = \frac{841\,\text{J}}{(95.6\,\text{g})(68.0°\text{C})} = 0.129\,\frac{\text{J}}{\text{g}°\text{C}}$$

5. What effect does the initial temperature of the water have on the change in temperature of the water after the hot metal is added? Explain your answer.

   No effect on Δt because the amount of water and the amount of energy added are what determine the change in temperature, not the initial temperature.

Results of scientific experiments must be reproducible when repeated or they do not mean anything. When results are repeated, the experiment is said to have good **precision**. When the results agree with a theoretical value they are described as **accurate**.

---

6. Which is better, the precision or the accuracy of the experimental specific heats determined for your metal sample? Support your answer with your data.

   (For sample data only) the accuracy is better because the average sp. ht. = 1.08 J/g°C is within 0.18 of the theoretical sp. ht. of 0.900 J/g°C. The 2 trials were different by 0.25 J/g°C so precision is not as good. (This answer may vary dependent on the data collected.)

Use the data presented in Table 5.1 to answer questions 7–9 and complete the graph on the next page to show the relationship between the atomic mass and the specific heat of the seven metals listed. Make the graph following the guidelines provided in Study Aid 3.

---

7. a. What is the independent variable? _____**atomic mass**_____

   b. What is the dependent variable? _____**specific heat**_____

8. In Table 5.1, what is the range of values for atomic mass? ____**26.98 to 207.2 = 180**____

9. In Table 5.1, what is the range of values for specific heat? ____**0.128 to 0.900 = 0.772**____

10. Plot the data in Table 5.1 on the graph below. Be sure to include the following:

    a. A title

    b. Placement of the independent and dependent variables on the appropriate axes

    c. Increments for each axis

    d. Labels for each axis

    e. Plotting the data points

> Typical Student graph of data in Table 5.1.

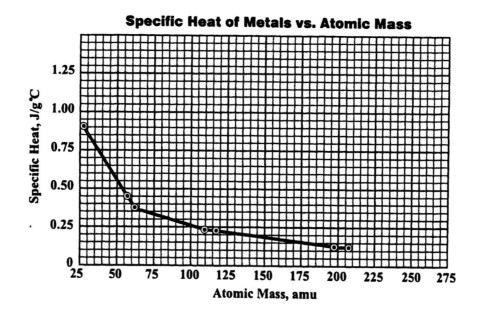

11. Use your graph to summarize the relationship between the atomic mass of metal atoms and the specific heat of a metal.

    As the atomic mass of a metal increases, its specific heat decreases (inverse relationship).

12. The specific heat was measured for two unknown metal samples. The first sample tested had a specific heat of 0.54 J/g°C. Use your graph to estimate the atomic mass of the metal.

    about 48 amu

    The second metal had a specific heat of 0.24 J/g°C. Use your graph again and estimate the atomic mass of this metal.

    about 100 amu

REPORT FOR EXPERIMENT 6

Freezing Points–Graphing of Data

> ✓ Student results will be variable and will not exactly match the typical student data shown on the key as a guideline for grading.

Data Table

time, minutes	Pure Acetic Acid temp, °C WITH STIRRING	Pure Acetic Acid temp, °C WITHOUT STIRRING	Impure Acetic Acid temp, °C WITHOUT STIRRING
0.0	25.0	25.0	25.0
0.5	(19.0)	11.0	16.0
1.0	17.0	9.0	14.0
1.5	17.0	7.5	11.0
2.0	17.0	7.0	10.0
2.5	17.0	6.0	8.0
3.0	17.0	5.5	7.5
3.5	17.0	5.0	7.0
4.0	17.0	5.0	6.5
4.5	17.0	5.0	6.0
5.0	17.0	(9.0)	6.0
5.5	17.0	16.0	6.0
6.0	17.0	16.0	6.0
6.5	17.0	16.0	5.5
7.0	17.0	16.0	5.5
7.5	17.0	16.0	5.0
8.0	17.0	16.0	(5.0)
8.5	17.0	16.0	12.0
9.0	17.0	16.0	12.5
9.5	17.0	16.0	12.5
10.0	17.0	16.0	12.5
10.5	17.0	16.0	12.5
11.0	17.0	16.0	12.5
11.5	17.0	16.0	12.5
12.0	17.0	16.0	12.5

Graphing of Freezing Point Data

Plot your data on the graph paper or the computer using a legend as follows:

△ = Pure acetic acid with stirring

▲ = Pure acetic acid without stirring

O = Acetic acid/benzoic acid solution without stirring

Draw rectangles around the portions of your curves that show supercooling.

QUESTIONS | ✓ These answers are based on the sample data and graph provided. Student data will be variable.

Use your graph to answer the questions 1–3.

1. a. At what temperature did crystals first form in Trial 1? _____ **19.0°C**

   b. Where did the temperature stabilize after supercooling in Trial 2? _____ **16.0°C**

   c. What is your experimental freezing point of glacial acetic acid? _____ **16.0 – 17.0°C**

   d. What is the theoretical freezing point of glacial acetic acid? _____ **16.6°C**
      (Consult the *Handbook of Chemistry and Physics*)

2. How many degrees was the freezing point depressed by the benzoic acid? _____ **3.5°C**

   Do this by estimating to the nearest 0.1 degree the number of degrees between the flattest (most nearly horizontal) portions of the curves. Mark the area on the graph with an arrow (↓) to show where this temperature difference estimate was made.

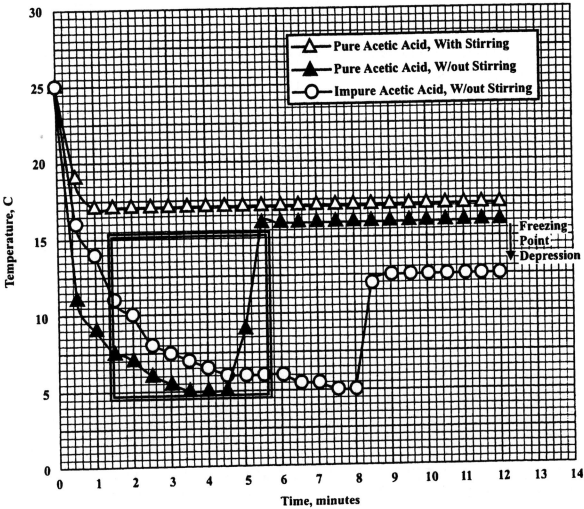

Freezing Point of Pure vs. Impure Acetic Acid With and Without Stirring

3. a. What is the effect of stirring on the freezing point of pure acetic acid?

   No effect. The freezing point was almost the same for the stirred and unstirred.

   b. What is the effect of stirring on supercooling?

   Stirring generally prevents supercooling.

4. a. What do the melting point and freezing point of a substance have in common?

   They are the same temperature.

   b. What is the difference between the melting and freezing of a substance?

   Heat is added to melt a solid substance; heat is removed to freeze a liquid substance.

5. When the solid and liquid phases are in equilibrium, which phase, solid or liquid contains the greater amount of energy? Explain the rationale for your answer.

   The liquid phase contains more energy. It is necessary to add heat (energy) to the solid phase to obtain the liquid phase.

NAME **KEY**

SECTION _____ DATE _____

REPORT FOR EXPERIMENT 7

INSTRUCTOR _____

Water in Hydrates

> ✓ Student results will be variable and will not exactly match the typical student data shown on the key as a guideline for grading.

A. Qualitative Determination of Water

1. Describe the appearance and odor of the liquid obtained by heating copper(II) sulfate pentahydrate.

   The liquid was colorless and had no noticeable odor.

2. Compare the results observed when testing the liquid from the hydrate and distilled water with the cobalt chloride paper and the anhydrous salt by completing the table below.

Property Observed	Cobalt Chloride Paper	Anhydrous $CuSO_4$
Color before adding liquid(s) to	**blue**	**white**
Color after adding distilled water to	**pink**	**blue**
Color after adding liquid from hydrate to	**pink**	**blue**
Temperature change after adding distilled water	**N/A**	**heat ↑**
Temperature change after adding liquid from hydrate	**N/A**	**heat ↑**

B. Quantitative Determination of Water in a Hydrate

1. Mass of crucible and cover .... 19.0280 g

2. Mass of crucible, cover, and sample .... 21.1811 g

3. Mass of crucible, cover, and sample after 1st heating ... 20.1403 g

4. Mass of crucible, cover, and sample after 2nd heating ... 20.1331 g

5. Mass of crucible, cover, and sample after 3rd heating (if needed) .... —

6. Mass of original sample .... 2.1531 g
   Show calculation setup:
$$\begin{array}{r} 21.1811\,g \\ -19.0280\,g \\ \hline 2.1531\,g \end{array}$$

7. Total mass lost by sample during heating .... 1.0480 g
   Show calculation setup: **21.1811 − 20.1331 = 1.0480**

8. Percentage water in sample $\left(\dfrac{1.0480\,g}{2.1531\,g}\right)(100)$ Sample No. _____ **A**
   Show calculation setup: 48.674%

– 65 –

QUESTIONS AND PROBLEMS

1. What evidence did you see that indicated the liquid obtained from the copper (II) sulfate pentahydrate was water?
   (a) **Reacted with $CoCl_2$ test paper in the same way as water.**

   (b) **Reacted with anhydrous salt in the same way as water.**

2. What was the evidence of a chemical reaction when the anhydrous salt samples were treated with the liquid obtained from the hydrate and with water?
   (a) **Both liquids caused the anhydrous salt to turn blue.**

   (b) **Heat was evolved (and steam and/or water vapor may have been observed).**

3. Write a balanced chemical equation for the decomposition of copper(II) sulfate pentahydrate.

   $$CuSO_4 \cdot 5\,H_2O \xrightarrow{\Delta} CuSO_4 + 5\,H_2O$$

4. When the unknown was heated, could the decrease in mass have been partly due to the loss of some substance other than water? Explain.

   Yes. Since we did not know what the samples were, it is possible that other or further decompositions may have occurred. (For example, $CuSO_4$ undergoes further decomposition.) $CuSO_4(s) \xrightarrow{\Delta} CuO(s) + SO_3(g)$

5. A student heated a hydrated salt sample with an initial mass of 4.8702 g, After the first heating, the mass had decreased to 3.0662 g.
   (a) If the sample was heated to constant weight after reheating, what is the minimum mass that the sample can have after the second weighing? Show how you determined your answer.

   $$\begin{array}{r} 3.0662 \\ -0.05 \\ \hline 3.0162\text{ g} \end{array}$$

   3.0162 g _____

   (b) The student determined that the mass lost by the sample was 1.8053. What was the percent water in the original hydrated sample? Show calculation setup.

   $$\left(\frac{1.8053\text{ g}}{4.8702\text{ g}}\right)(100) = 37.068\%$$

   37.068% _____

REPORT FOR EXPERIMENT 8

Properties of Solutions

> ✓ Student results will be variable and will not exactly match the typical student data shown on the key as a guideline for grading.

A. Concentration of Saturated Solution

1. Mass of empty evaporating dish _____ 69.1476 g

2. Mass of dish + saturated potassium chloride solution _____ 75.9654 g

3. Mass of dish + dry potassium chloride, 1st heating _____ 71.0084 g

4. Mass of dish + dry potassium chloride, 2nd heating _____ 71.0054 g

5. Mass of saturated potassium chloride solution _____ 6.8178 g
   Show Calculation Setup
   $$75.9654$$
   $$-69.1476$$

6. Mass of potassium chloride in the saturated solution _____ 1.8578 g
   Show Calculation Setup
   $$71.0054$$
   $$-69.1476$$

7. Mass of water in the saturated potassium chloride solution _____ 4.9600 g
   Show Calculation Setup
   $$75.9654$$
   $$71.0054$$

8. Mass percent of potassium chloride in the saturated solution _____ 27.249%
   Show Calculation Setup   **(range 23.0 − 28.0%)**
   $$\left(\frac{1.8578\,g}{4.9600 + 1.8578}\right)(100) = \left(\frac{1.8578\,g}{6.8178\,g}\right)(100) =$$

9. Grams of potassium chloride per 100 g of water (experimental) $37.456\,g\,KCl/100\,g\,H_2O$ in the original solution.
   Show Calculation Setup

   $$\left(\frac{1.8578\,g\,KCl}{4.9600\,H_2O}\right)(100)$$

10. Grams of potassium chloride per 100 g of water (theoretical) $34.0\,g\,KCl/100\,g\,H_2O$ (From Table 8.1) at 20°C.

B. Relative Solubility of a Solute in Two Solvents

1. (a) Which liquid is denser, decane or water?   ____**Water**____

    (b) What experimental evidence supports your answer?

    Decane is the smaller of the two volumes measured. It appears as the top layer and therefore is less dense than water which went to the bottom.

2. Color of iodine in water: **yellow-brown**

    Color of iodine in decane: **bright pink/violet**

3. (a) In which of the two solvents used is iodine more soluble?   **decane**

    (b) Cite experimental evidence for your answer.

    After shaking the color of iodine in the water layer was very light and the color of decane was bright pink. (Showing a transfer of I_2 from the water to decane.)

C. Miscibility of Liquids

1. Which liquid pairs tested are miscible?

    Kerosene and isopropyl alcohol
    Water and isopropyl alcohol

2. How do you classify the liquid pair decane—H_2O, miscible or immiscible?

                                                                        ____**immiscible**____

D. Rate of Dissolving Versus Particle Size

1. Time required for fine salt crystals to dissolve   **about 30 sec**

2. Time required for coarse salt crystals to dissolve   **about 2 min**
                                                                  (greater than above)

3. Since the amount of salt, the volume of water, and the temperature of the systems were identical in both test tubes, how do you explain the difference in time for dissolving the fine vs. the coarse salt crystals?

    Small crystals or particles dissolve faster than large ones.

E. Rate of Dissolving Versus Temperature

1. Under which condition, hot or cold, did the salt dissolve faster?   **hot**

2. Since the amount of salt, the volume of water, and the texture of the salt crystals were identical in both best tubes, how do you explain the difference in time for dissolving at the hot vs. cold temperatures?

    Solvent (water) particles are moving faster at higher temperatures and collide with solute particles (salt) more often causing the rate of dissolving to increase.

F.   Solubility vs. Temperature; Saturated and Unsaturated Solutions

Data Table: Circle the choices which best describe your observations.

	NaCl	NH₄Cl
1.0 g + 5 mL water	dissolved completely? (yes)/no saturated or (unsaturated?)	dissolved completely? (yes)/no saturated or (unsaturated?)
1.0 g + 5 mL water + 1.4 g	dissolved completely? yes/(no) (saturated) or unsaturated?	dissolved completely? yes/(no) (saturated) or unsaturated?
2.4 g + 5 mL water + heat	dissolved completely? yes/(no) (saturated) or unsaturated?	dissolved completely? (yes)/no saturated or (unsaturated?)
2.4 g + 5 mL water after cooling	dissolved completely? yes/(no) (saturated) or unsaturated?	dissolved completely? yes/(no) (saturated) or unsaturated?

G.   Ionic Reactions in Solution

1. Write the word and formula equations representing the chemical reaction that occurred between the barium chloride solution, $BaCl_2(aq)$, and the sodium sulfate solution, $Na_2SO_4(aq)$.

   Word Equation:

   Barium chloride + Sodium sulfate $\longrightarrow$ Barium sulfate + Sodium chloride

   Formula Equation:

   $BaCl_2(aq) + Na_2SO_4(aq) \longrightarrow BaSO_4(s) + 2\,NaCl(aq)$

2. (a) Which of the products is the white precipitate?   ___**BaSO₄**___

   (b) What experimental evidence leads you to this conclusion?

   Of the four compounds tested for solubility, BaSO₄ is the only one that was insoluble.

SUPPLEMENTARY QUESTIONS AND PROBLEMS

1. Use the solubility data in Table 8.1 in answering the following:
   Show Calculations

   (a) What is the percentage by mass of NaCl in a saturated solution of sodium chloride at 50°C?   **37.0 g NaCl + 100. g H₂O = 137 g total mass**

   $$\left(\frac{37.0\,g}{137\,g}\right)(100) = 27.0\%\ NaCl$$ ___**27.0% NaCl**___

(b) Calculate the solubility of potassium bromide at 23°C. Hint: Assume that the solubility increases by an equal amount for each degree between 20°C and 30°C.

70.6 g @ 30°C $\underline{66.8\,g/100\,g\,H_2O}$

65.2 g @ 20°C

   5.4 g increase in solubility from 20°C to 30°C

Sol. @ 23° = 65.2 g + 0.3(5.4) g = 66.8 g/100 g H_2O

(c) A saturated solution of barium chloride at 30°C contains 150 g water. How much additional barium chloride can be dissolved by heating this solution to 60°C?

@ 30°C $150\,g\,H_2O \times \dfrac{38.2\,g\,BaCl_2}{100\,g\,H_2O} = 57.3\,g\,BaCl_2$ $\underline{\quad 12.6\,g \quad}$

@ 60°C $150\,g\,H_2O \times \dfrac{46.6\,g\,BaCl_2}{100\,g\,H_2O} = 69.9\,g\,BaCl_2$

69.9 − 57.3 g = 12.6 g

2. A solution of KCl is saturated at 50°C.
   Use Table 8.1

   (a) How many grams of solute are dissolved in 100 g of water? $\underline{\quad 42.6\,g \quad}$

   from table 8.1

   (b) What is the total mass of the solution? $\underline{\quad 142.6\,g \quad}$

   100 g water + 42.6 g KCl

   (c) What is the mass percent of this solution at 50°C? $\underline{\quad 29.9\% \quad}$

   $\left(\dfrac{42.6\,g}{142.6\,g}\right)(100) = 29.9\%$

   (d) If the solution is heated to 100°C, how much more KCl can be dissolved in the solution without adding more water?

   at 100°   55.6 g − 42.6 g = $\underline{\quad 13.0\,g \quad}$

   (e) If the solution is saturated at 100°C and then cooled to 30°C, how many grams of solute will precipitate out?

   55.6 g @ 100°C $\underline{\quad 18.6\,g \quad}$

   37.0 g @ 30°C

   18.6 g

NAME **KEY**

SECTION _____ DATE _____

REPORT FOR EXPERIMENT 9

INSTRUCTOR _____

Composition of Potassium Chlorate

✓ Student results will be variable and will not exactly match the typical student data shown on the key as a guideline for grading.

A. Determining Percentage Composition	Sample 1	Sample 2
1. Mass of crucible + cover	20.1443 g	_____
2. Mass of crucible + cover + sample before heating	21.4095 g	_____
3. Mass of crucible + cover + residue after 1st heating	20.9762 g	_____
4. Mass of crucible + cover + residue after 2nd heating	20.9184 g	_____
5. Mass of crucible + cover + residue after 3rd heating (if necessary)	20.9180 g	_____

6. Mass of original sample

Show sample 1 calculation setup:

$$\begin{array}{r} 21.4095 \\ -20.1443 \\ \hline 1.2652 \end{array}$$

    1.2652 g   _____

7. Mass lost (total) during heating

Show sample 1 calculation setup:

21.4095 g − 20.9180 g = 0.4915 g

    0.4915 g   _____

8. Final mass of residue

Show sample 1 calculation setup:

20.9180 g − 20.1443 g = 0.7737 g

    0.7737 g   _____

9. Experimental percent oxygen in sample ($KClO_3$)

Show sample 1 calculation setup: $\left(\dfrac{0.49159\ g}{1.2652\ g}\right)(100) = 38.85$

    38.85%   _____

10. Experimental percent KCl in sample ($KClO_3$)

Show sample 1 calculation setup: $\left(\dfrac{0.7737\ g}{1.2652\ g}\right)(100) = 61.15\%$

    61.15%   _____

11. Theoretical percent oxygen in $KClO_3$

Show calculation setup:

$$\begin{array}{lll} K & 1 \times & 39.10 \\ Cl & 1 \times & 35.45 \\ O & 3 \times & 16.00 \\ \hline & & 122.6\ g \end{array}$$

    39.15%

12. Theoretical percent KCl in $KClO_3$

Show calculation setup

$\left(\dfrac{39.10 + 35.45g}{122.6\ g}\right)(100) =$

    60.81%

13. Percent error in experimental % oxygen determination

Show sample 1 calculation setup

$\dfrac{theoretical - experimental}{theoretical}(100)$ $\dfrac{39.15\% - 38.85\%}{39.15\%}(100)$

    0.7663%   _____

- 81 -

B.   Qualitative Examination of Residue

1.   Record what you observed when silver nitrate was added to the following:

(a) Potassium chloride solution   **white precipitate formed**

(b) Potassium chlorate solution   **no precipitate formed <u>or</u> very slight precipitate (cloudiness) formed.**

(c) Residue solution   **white precipitate formed**

2.   (a) What evidence did you observe that would lead you to believe that the residue was potassium chloride?
Residue and potassium chloride reacted similarly (formed ppt) with AgNO$_3$.

(b) What would happen if you added silver nitrate to a solution of sodium chloride? Explain your answer.
A precipitate would form because the silver nitrate reacts with the chloride ion.

(c) Did the evidence obtained in the silver nitrate tests of the three solutions prove conclusively that the residue actually was potassium chloride? Explain?
No. The residue could have been any soluble chloride and the results would be the same.

QUESTIONS AND PROBLEMS

1.   A student forgot to read the label on the jar carefully and put potassium chloride in the crucible instead of potassium chlorate. How would the results turn out?
The sample would not lose any mass when heated.

2.   What if a potassium chlorate sample is contaminated with KCl. Would the experimental % oxygen be higher or lower than the theoretical % oxygen? Explain your answer.
% oxygen would be lower because the total mass would include the KCl which does not lose any oxygen when heated.

3.   What if a potassium chlorate sample is contaminated with moisture. Would an analysis show the experimental % oxygen higher or lower than the theoretical % oxygen? Explain your answer?
% oxygen would be higher because the moisture would be lost during heating in addition to the oxygen.

4.   Calculate the percentage of Cl in Al(ClO$_3$)$_3$

$$\%Cl = \left(\frac{3Cl}{Al(ClO_3)_3}\right)(100) = \frac{3(35.45)}{277.3 \text{ g}}(100) =$$

_____**38.35%**_____

5.   Other metal chlorates when heated show behavior similar to that of potassium chlorate yielding metal chlorides and oxygen. Write the balanced formula equation for the reaction to be expected when calcium chlorate, Ca(ClO$_3$)$_2$ is heated.

Ca(ClO$_3$)$_2$ $\xrightarrow{\Delta}$ CaCl$_2$ + 3O$_2$(g)

REPORT FOR EXPERIMENT 10

Double Displacement Reactions

Directions for completing table below:

1. Record your observations (Evidence of Reaction) of each experiment. Use the following terminology: (a) "Precipitate formed" (include the color), (b) "Gas evolved," (c) "Heat evolved," or (d) "No reaction observed."

2. Complete and balance the equation for each case in which a reaction occurred. First write the correct formulas for the products, taking into account the charges (oxidation numbers) of the ions involved. Then balance the equation by placing a whole number in front of each formula (as needed) to adjust the number of atoms of each element so that they are the same on both sides of the equation. Use (g) or (s) to indicate gases and precipitates. Where no evidence of reaction was observed, write the words "No reaction" as the right-hand side of the equation.

Evidence of Reaction	Equation		
1. No reaction observed	$NaCl$ +	$KNO_3 \longrightarrow$	No reaction
2. White precipitate	$NaCl$ +	$AgNO_3 \longrightarrow$	$AgCl(s)$ + $NaNO_3$
3. Gas evolved	Na_2CO_3 +	$2\,HCl \longrightarrow$	$2\,NaCl$ + H_2O + $CO_2(g)$
4. Heat evolved	$NaOH$ +	$HCl \longrightarrow$	$NaCl$ + H_2O
5. White precipitate formed	$BaCl_2$ +	$H_2SO_4 \longrightarrow$	$BaSO_4(s)$ + $2\,HCl$
6. Heat evolved	$2\,NH_4OH$ +	$H_2SO_4 \longrightarrow$	$(NH_4)_2SO_4$ + $2\,H_2O$*
7. No reaction observed	$CuSO_4$ +	$Zn(NO_3)_2 \longrightarrow$	No reaction
8. White precipitate formed	Na_2CO_3 +	$CaCl_2 \longrightarrow$	$CaCO_3(g)$ + $2\,NaCl$
9. No reaction observed	$CuSO_4$ +	$NH_4Cl \longrightarrow$	No reaction
10. Heat evolved	$NaOH$ +	$HNO_3 \longrightarrow$	$NaNO_3$ + H_2O
11. Reddish precipitate formed (orange or brown)	$FeCl_3$ +	$3\,NH_4OH \longrightarrow$	$Fe(OH)_3(s)$ + $3\,NH_4Cl$
12. Gas evolved	Na_2SO_3 +	$2\,HCl \longrightarrow$	$2\,NaCl$ + H_2O + $SO_2(g)$

*or NH_4OH + $H_2SO_4 \longrightarrow NH_4HSO_4$ + H_2O

QUESTIONS AND PROBLEMS

1. The formation of what three classes of substances caused double displacement reactions to occur in this experiment?

   (a) **Precipitates**

   (b) **Gases**

   (c) **Slightly ionized substances (usually water)**

2. Write the equation for the decomposition of sulfurous acid.

   $$H_2SO_3 \longrightarrow H_2O + SO_2(g)$$

3. Using three criteria for double displacement reactions, together with the Solubility Table in Appendix 5, predict whether a double displacement reaction will occur in each example below. If reaction will occur, complete and balance the equation, properly indicating gases and precipitates. If you believe no reaction will occur, write "no reaction" as the right-hand side of the equation. All reactants are in aqueous solution.

   (a) K_2S + $CuSO_4 \longrightarrow$ $CuS(s) + K_2SO_4$

   (b) 2 NH_4OH + $H_2C_2O_4 \longrightarrow$ $(NH_4)_2C_2O_4 + 2\,H_2O$

   (c) KOH + $NH_4Cl \xrightarrow{\Delta}$ $KCl + H_2O + NH_3(g)$

   (d) $NaC_2H_3O_2$ + $HCl \longrightarrow$ $NaCl + HC_2H_3O_2$

   (e) Na_2CrO_4 + $Pb(C_2H_3O_2)_2 \longrightarrow PbCrO_4(s) + 2\,NaC_2H_3O_2$

   (f) $(NH_4)_2SO_4$ + $NaCl \longrightarrow$ **No reaction**

   (g) $BiCl_3$ +   3 $NaOH \longrightarrow$ $Bi(OH)_3(s) + 3\,NaCl$

   (h) $KC_2H_3O_2$ + $CoSO_4 \longrightarrow$ **No reaction**

   (i) Na_2CO_3 +   2 $HNO_3 \longrightarrow$ $2\,NaNO_3 + H_2O + CO_2(g)$

   (j) 3 $ZnBr_2$ +   2 $K_3PO_4 \longrightarrow$ $Zn_3(PO_4)_3(s) + 6\,KBr$

REPORT FOR EXPERIMENT 11

INSTRUCTOR _____

Single Displacement Reactions

Evidence of Reaction	Equation (to be completed)
Describe any evidence of reaction; if no reaction was observed, write "None".	Write "No reaction", if no reaction was observed.
1. **Deposit on Cu strip (and sol'n. develops blue color)**	$Cu +$ $2\,AgNO_3(aq) \longrightarrow$ $2\,Ag(s) + Cu(NO_3)_2$
2. **Deposit on Pb strip**	$Pb +$ $Cu(NO_3)_2(aq) \longrightarrow Cu(s) + Pb(NO_3)_2$
3. **Deposit on Zn strip**	$Zn +$ $Pb(NO_3)_2(aq) \longrightarrow Pb(s) + Zn(NO_3)_2$
4. **None**	$Zn +$ $MgSO_4(aq) \longrightarrow$ **No reaction**
5. **None**	$Cu +$ $H_2SO_4(aq) \longrightarrow$ **No reaction**
6. **Gas evolved (and metal gradually disappeared)**	$Zn +$ $H_2SO_4(aq) \longrightarrow$ $H_2(g) + ZnSO_4$

QUESTIONS AND PROBLEMS

1. Complete the following table by writing the symbols of the two elements whose reactivities are being compared in each test:

	Tube Number					
	1	**2**	**3**	**4**	**5**	**6**
Greater activity	**Cu**	**Pb**	**Zn**	**Mg**	**H**	**Zn**
Lesser activity	**Ag**	**Cu**	**Pb**	**Zn**	**Cu**	**H**

2. Arrange Pb, Mg, and Zn in order of their activities, listing the most active first.

(1) _____ **Mg** _____

(2) _____ **Zn** _____

(3) _____ **Pb** _____

3. Arrange Cu, Ag, and Zn in order of their activities, listing the most active first.

(1) _____ **Zn** _____

(2) _____ **Cu** _____

(3) _____ **Ag** _____

4. Arrange Mg, H, and Ag in order of their activities, listing the most active first.

(1) _____ **Mg** _____

(2) _____ **H** _____

(3) _____ **Ag** _____

5. Arrange all five of the metals (excluding hydrogen) in an activity series, listing the most active first.

(1) _____ **Mg** _____

(2) _____ **Zn** _____

(3) _____ **Pb** _____

(4) _____ **Cu** _____

(5) _____ **Ag** _____

6. On the basis of the reactions observed in the six test tubes, explain why the position of hydrogen cannot be fixed exactly with respect to all of the other elements listed in the activity series in Question 5.

Hydrogen is more reactive than Cu and less reactive than Zn, but its reactivity has not been established with respect to Pb.

7. What additional test(s) would be needed to establish the exact position of hydrogen in the activity series of the elements listed in Question 5?

It would be necessary to determine the relative activity of Pb and H.

8. On the basis of the evidence developed in this experiment:

(a) Would silver react with dilute sulfuric acid? Why or why not?

No. Since silver is less reactive than Cu and Cu does not react with dil. H_2SO_4, Ag will not react with dil. H_2SO_4, or No, since silver is less reactive than hydrogen.

(b) Would magnesium react with dilute sulfuric acid? Why or why not?

Yes. Since Mg is more reactive than Zn and Zn reacts with dil. H_2SO_4, Mg will react with dil. H_2SO_4, or Yes, since magnesium is more reactive than hydrogen.

REPORT FOR EXPERIMENT 12

Ionization—Electrolytes and pH

A. Conductivity of Solutions—Instructor Demonstration

Complete the table for each of the substances tested in the ionization demonstration. Place an "X" in the column where the property of the substance tested fits the column description.

	Nonelectrolyte	Strong Electrolyte	Weak Electrolyte
1. Distilled Water	X		
2. Tap water			X
3. Sugar	X		
4. NaCl		X	
5. a. $HC_2H_3O_2$ (glacial)	X		
b. 1st dilution			X
c. 2nd dilution			X
6. a. 1 M $HC_2H_3O_2$			X
b. 1 M HCl		X	
c. 1 M NH_4OH			X
d. 1 M NaOH		X	
7. a. $NaNO_3$		X	
b. NaBr		X	
c. $Ni(NO_3)_2$		X	
d. $CuSO_4$		X	
e. NH_4Cl		X	

8. (a) Write an equation for the chemical reaction that occurred between sulfuric acid and barium hydroxide.

$$H_2SO_4 + Ba(OH)_2 \longrightarrow BaSO_4(s) + 2\,H_2O$$

   (b) Explain in terms of the properties of the products formed why the light went out when barium hydroxide was added to sulfuric acid solution, even though both of these reactants are electrolytes.

   The ions from $Ba(OH)_2$ reacted with those from H_2SO_4 to form $BaSO_4$, an insoluble salt, and H_2O, a nonelectrolyte. The lights went out when there were not enough ions left in solution to conduct the electric current.

   (c) Explain why the light came on again when additional barium hydroxide was added.

   The light came on again because ions from the additional $Ba(OH)_2$ conducted the electric current.

9. In the conductivity tests, what controlled the brightness of the light?

   The number of ions in solution (the depth to which the electrodes are immersed in the solution).

10. Write an equation to show how acetic acid reacts with water to produce ions in solution.

$$HC_2H_3O_2 + H_2O \longrightarrow H_3O^+ + C_2H_3O_2^- \quad \text{or} \quad HC_2H_3O_2 \longrightarrow H^+ + C_2H_3O_2^-$$

11. What classes of compounds tested are electrolytes?

   Acids, bases, and salts.

B. Properties of Acids

1. Reaction with a Metal

   (a) Write the formulas of the acids which liberated hydrogen gas when reacting with magnesium metal.

   _____**$HCl, H_2SO_4,$ and $HC_2H_3O_2$**_____

   (b) Write equations to represent the reactions in which hydrogen gas was formed.

$$Mg + 2\,HCl \longrightarrow MgCl_2 + H_2(g)$$

$$Mg + H_2SO_4 \longrightarrow MgSO_4 + H_2(g)$$

$$Mg + 2\,HC_2H_3O_2 \longrightarrow Mg(C_2H_3O_2)_2 + H_2(g)$$

2. Measurement of Acidity and pH

(a) What is the effect of acids on the color of red litmus?

No color change.

(b) What is the effect of acids on the color of blue litmus?

Acids change the color of blue litmus to red.

(c) What color is phenolphthalein in an acid solution?   **Colorless**

(d) What was the pH of the hydrochloric acids tested?

0.001 M __**about 3**__   0.01 M __**about 2**__   0.1 M __**about 1**__

(e) Which pH measured has the highest number of H^+ in solution?   **pH = 1**

(f) What is the H^+ concentration in an acid with of pH 4.6?
Express your answer as a power of 10   **2.51×10^{-5} M**

Refer to Study Aid 4 if you need help with using your calculator to convert the pH into H^+ concentration using the antilog function.

3. Reaction with Carbonates and Bicarbonates

(a) What gas is formed in these reactions?

Name _____**Carbon dioxide**_____   Formula _____**CO_2**_____

(b) What happened to the burning splint when it was thrust into the beaker?

It was extinguished.

(c) What do you conclude about one of the properties of the gas in the beaker, based on the behavior of the burning splint?

The gas (CO_2) does not support combustion.

(d) Complete and balance the equations representing the reactions:

$NaHCO_3(s) + HCl(aq) \longrightarrow NaCl + H_2O + CO_2(g)$

$CaCO_3(s) + 2\,HCl(aq) \longrightarrow CaCl_2 + H_2O + CO_2(g)$

4. Reaction of Acids with Bases—Neutralization

(a) Write an equation for the neutralization reaction of HCl and NaOH.

HCl + NaOH $\longrightarrow$ NaCl + H$_2$O

(b) How did you know when all the acid was neutralized?

The indicator (phenolphthalein) in the solution changes to a pink color.

5. Nonmetal Oxide plus Water

(a) Write an equation for the combustion of sulfur in air.

S + O$_2$ $\longrightarrow$ SO$_2$

(b) What acid is formed when the product of the sulfur combustion reacts with water?

Name _____**Sulfurous acid**_____ Formula _____**H$_2$SO$_3$(aq)**_____

Write the equation for its formation.

SO$_2$ + H$_2$O $\longrightarrow$ H$_2$SO$_3$(aq)

(c) What evidence in this experiment leads you to believe that carbon dioxide in water has acidic properties?

The pink color of phenolphthalein changed to colorless when CO$_2$ was bubbled into the alkaline solution.

(d) What acid is formed when carbon dioxide reacts with water?

Name _____**Carbonic acid**_____ Formula _____**H$_2$CO$_3$(aq)**_____

C. Properties of Bases

1. "Feel" Test. What is the characteristic feel of basic solutions?

slippery or soapy or oily

2. Measurement of Alkalinity

(a) What is the effect of bases on the color of red litmus?

Bases cause the color of red litmus to turn blue.

(b) What is the effect of bases on the color of blue litmus?

   No change in color.

(c) What color is phenolphthalein in a basic solution?   ____**pink**____

(d) What was the pH for each dilute base tested?

   $NH_4OH(aq)$ ___**about 8**___ $NaOH(aq)$ ___**about 11**___

(e) Which base tested has the highest number of H^+ in solution?   ____**NH₄OH**____

(f) What is the H^+ concentration in the strongest base tested? Express your answer as a power of 10.   ____**10^{-11} M**____

   Refer to Study Aid 4 if you need help using your calculator to convert the pH into H^+ concentration using the antilog function.

3. **Metal Oxides plus Water**

   (a.1) Color (if any) produced by phenolphthalein.

   Color with CaO in water   ____**pink**____

   Color with MgO in water   ____**pink**____

   Color with $Ca(OH)_2$ in water   ____**pink**____

   (a.2) Complete and balance these equations:

   $$CaO + H_2O \longrightarrow Ca(OH)_2$$

   $$MgO + H_2O \longrightarrow Mg(OH)_2$$

   (b.1) The formula for marble is $CaCO_3$, What compounds are formed when it is heated strongly?

   CaO and CO₂

   (b.2) Write the equation representing this decomposition:

   $$CaCO_3(s) \xrightarrow{\Delta} CaO(s) + CO_2(g)$$

   (b.3) What evidence led you to formulate the composition of the solid residue after heating the marble chip?

   When the residue was dropped into the water containing phenolphthalein, the solution became pink in color. This same color change occurred with known samples of CaO and Ca(OH)₂. There was no color change with the unheated marble chip.

ADDITIONAL QUESTIONS AND PROBLEMS

1. State whether each of the formulas below represents an **acid**, a **base**, a **salt**, an **acid anhydride**, a **basic anhydride**, or **none** of these types of compounds:

CuF_2	salt	$CaSO_4$	salt
$Ba(OH)_2$	base	C_2H_4	none of these types
$LiOH$	base	$C_{12}H_{22}O_{11}$	none of these types
$HBrO_3$	acid	HI	acid
$RaCO_3$	salt	P_2O_5	none (acid anhydride)
KNO_2	salt	HCN	acid
$H_2C_2O_4$	acid	MgO	basic anhydride

2. Complete and balance the following equations and name the product formed. (Only one product is formed in each case.)

Name of Product

(a) $K_2O(s)$ + $H_2O(l) \longrightarrow$ **2 KOH**   <u>potassium hydroxide</u>

(b) $SrO(s)$ + $H_2O(l) \longrightarrow$ **Sr(OH)₂**   <u>strontium hydroxide</u>

(c) $SO_3(s)$ + $H_2O(l) \longrightarrow$ **H₂SO₄**   <u>sulfuric acid</u>

(d) $N_2O_5(s)$ + $H_2O(l) \longrightarrow$ **2 HNO₃**   <u>nitric acid</u>

REPORT FOR EXPERIMENT 13

Identification of Selected Anions

	NaCl	NaBr	NaI	Na$_2$SO$_4$	Na$_3$PO$_4$	Na$_2$CO$_3$	Unknown No. ___	Unknown No. ___
A. AgNO$_3$ Test Addition of AgNO$_3$ solution	white ppt. formed	pale yellow ppt. formed	yellow ppt.	no ppt.	yellow ppt.	yellow to brn. ppt.		
Addition of dil. HNO$_3$	ppt. did not dissolve	ppt. did not dissolve	ppt. did not dissolve	—	ppt. dissolved	ppt. dissolved		
B. BaCl$_2$ Test Addition of BaCl$_2$ solution	no ppt.	no ppt.	no ppt.	white ppt.	white ppt.	white ppt.		
Addition of dil. HCl	—	—	—	ppt. didn't dissolve	ppt. dissolved	ppt. dissolved		
C. Organic Solvent Test Color of decane layer	colorless to pale yellow	yellow-orange to red brown	pink to violet	colorless or pale yellow	colorless or pale yellow	colorless or pale yellow		
D. Formula of anion present in the solution tested.	Cl$^-$	Br$^-$	I$^-$	SO$_4^{2-}$	PO$_4^{3-}$	CO$_3^{2-}$		

– 111 –

QUESTIONS AND PROBLEMS

1. The following three solutions were analyzed according to the scheme used in this experiment. Which one, if any, of the ions tested, is present in each solution? If the data indicate that none of the six is present, write the word "None" as your answer.

   (a) **Silver Nitrate Test.**   Yellow precipitate formed, which dissolved in dilute nitric acid.

   Barium Chloride Test.   White precipitate formed, which dissolved in dilute hydrochloric acid.

   Organic Solvent Test.   The decane layer remained almost colorless after treatment with chlorine water.

   Anion present _____ PO_4^{3-} _____

   (b) **Silver Nitrate Test.**   Red precipitate formed, which dissolved in dilute nitric acid to give an orange solution.

   Barium Chloride Test.   Yellow precipitate formed, which dissolved in dilute hydrochloric acid to give an orange solution.

   Organic Solvent Test.   The decane layer remained almost colorless after treatment with chlorine water.

   Anion present _____ **none** _____

   (c) **Silver Nitrate Test.**   Yellow precipitate formed, which did not dissolve in dilute nitric acid.

   Barium Chloride Test.   No precipitate formed.

   Organic Solvent Test.   The decane layer turned reddish-brown.

   Anion present _____ Br^- _____

2. Write un-ionized, total ionic, and net ionic equations for the following reactions: Use the solubility table in Appendix 5 for reactions that were not observed directly in this experiment.

(a) Sodium bromide and silver nitrate.

$$NaBr + AgNO_3 \longrightarrow AgBr(s) + NaNO_3$$

$$Na^+ + Br^- + Ag^+ + NO_3^- \longrightarrow AgBr(s) + Na + NO_3^-$$

$$Br^- + Ag^+ \longrightarrow AgBr(s)$$

(b) Sodium carbonate and silver nitrate.

$$Na_2CO_3 + 2\,AgNO_3 \longrightarrow Ag_2CO_3(s) + 2\,NaNO_3$$

$$2\,Na^+ + CO_3^{2-} + 2\,Ag^+ + 2\,NO_3^- \longrightarrow Ag_2CO_3(s) + 2\,Na + 2\,NO_3^-$$

$$CO_3^{2-} + 2\,Ag^+ \longrightarrow Ag_2CO_3(s)$$

(c) Sodium arsenate and barium chloride.

$$2\,Na_3AsO_4 + 3\,BaCl_2 \longrightarrow Ba_3(AsO_4)_2(s) + 6\,NaCl$$

$$6\,Na^+ + 2\,AsO_4^{3-} + 3\,Ba^{2+} + 6\,Cl^- \longrightarrow Ba_3(AsO_4)_2(s) + 6\,Na^+ + 6\,Cl^-$$

$$2\,AsO_4^{3-} + 3\,Ba^{2+} \longrightarrow Ba_3(AsO_4)_2(s)$$

3. Write net ionic equations for the following reactions. Assume that a precipitate is formed in each case.

(a) Sodium iodide and silver nitrate.

$$I + Ag^+ \longrightarrow AgI(s)$$

(b) Sodium acetate and silver nitrate.

$$C_2H_3O_2^- + Ag^+ \longrightarrow AgC_2H_3O_2(s)$$

(c) Sodium phosphate and barium chloride.

$$2\,PO_4^{3-} + 3\,Ba^{2+} \longrightarrow Ba_3(PO_4)_2(s)$$

(d) Sodium sulfate and barium chloride.

$$SO_4^{2-} + Ba^{2+} \longrightarrow BaSO_4(s)$$

NAME _____ **KEY** _____

SECTION _____ DATE _____

REPORT FOR EXPERIMENT 14

INSTRUCTOR _____

Properties of Lead(II), Silver and Mercury(I) Ions

A. Test for Lead(II) Ion

  1. Record your observations for

   (a) Part A.1.

   A white precipitate formed.

   (b) Part A.4.

   The precipitate dissolved.

   (c) Part A.5.

   A yellow precipitate formed.

  2. Write the name, formula, and color of the precipitate formed in Part A.1.

   Lead (II) chloride
   _____

   $PbCl_2$
   _____

   White
   _____

  3. Write the name, formula, and color of the precipitate formed in the confirmatory test for lead(II) ion (Part A.5).

   Lead (II) iodide
   _____

   PbI_2
   _____

   Yellow
   _____

  4. What is accomplished by washing a precipitate?

   Washing removes soluble ions remaining on the precipitate so they will not interfere with subsequent tests.

B. Test for Silver Ion

1. Record your observations for

    (a) Part B.1.

    A white precipitate formed.

    (b) Part B.4.

    The precipitate dissolved.

    (c) Part B.5.

    A white precipitate formed.

2. Write the name, formula, and color of the precipitate formed in Part B.1.

Silver chloride
AgCl
White

3. Write the name, formula, and color of the precipitate formed in the confirmatory test for silver ion (Part B.5.).

Silver chloride
AgCl
White

C. Tests for Mercury(I) Ion

1. Record your observations for

    (a) Part C.1.

    A white precipitate formed.

    (b) Part C.4.

    The precipitate turned black.

2. Write the name, formula, and color of the precipitate formed in Part C.1.

Mercury (I) chloride

Hg_2Cl_2

White

3. Write the name, formula, and color of the two substances formed in the confirmatory test for mercury(I) ion (Part C.4).

Mercury	**Mercury (II) amido chloride**
Hg	**$HgNH_2Cl$**
Black	**White**

D. Analysis of a Known and an Unknown Solution

1. The following questions pertain to the **known solution**.

   (a) Write the formulas of the substances precipitated when HCl was added (Part D.1).

   $PbCl_2, AgCl, Hg_2Cl_2$

   (b) What occurred when this precipitate was treated with hot water? What is the evidence for your conclusion (Part D.4)?

   Some of the precipitate ($PbCl_2$) dissolved. Evidence: When the hot water filtrate was tested with NaI, a yellow precipitate formed <u>or</u> the volume of the original precipitate appeared smaller (diminished).

   (c) After filtering (Part D.5), why was the precipitate washed with more hot water?

   To remove any remaining $PbCl_2$.

   (d) What two reactions occurred simultaneously to the precipitate in Part D.8 when concentrated NH_4OH was added after the hot water wash?

   (1) Part of the precipitate dissolved <u>or</u> AgCl dissolved.
   (2) Part of the precipitate turned black <u>or</u> Hg_2Cl_2 turned black.

   (e) What did you observe when you added dilute HNO_3 to the filtrate in Part D.9?

   A white precipitate formed.

2. Unknown No. _____   Cations present _____

QUESTIONS AND PROBLEMS

1. Suggest another reagent that could be used in place of hydrochloric acid to precipitate the cations of the silver group and still allow a sample to be analyzed by the scheme used in this experiment.

   NaCl or any other soluble chloride.

2. For each of the following pairs of chlorides, select a reagent that will dissolve one of them and thus allow the separation of the two compounds.

   $AgCl - PbCl_2$ **Hot water or NH_4OH**

   $AgCl - Hg_2Cl_2$ **NH_4OH**

   $PbCl_2 - Hg_2Cl_2$ **Hot water**

3. What conclusions can be drawn about the cation(s) present in silver group unknowns that showed the following characteristics? Use formulas for the ions present.

   (a) A white chloride precipitate was partially soluble in hot water and turned black when concentrated ammonium hydroxide was added to it.

   Cation(s) Present   **Pb^{2+} and Hg_2^{2+}**

   (b) A white precipitate was formed on addition of hydrochloric acid. The precipitate was insoluble in hot water and soluble in concentrated ammonium hydroxide.

   Cation(s) Present   **Ag^+**

   (c) A white precipitate was formed on addition of hydrochloric acid and it dissolved when the solution was heated.

   Cation(s) Present   **Pb^{2+}**

NAME _____ **KEY** _____

SECTION _____ DATE _____

REPORT FOR EXPERIMENT 15   INSTRUCTOR _____

> ✓ Student results will be variable and will not exactly match the typical student data shown on the key as a guideline for grading.

Quantitative Preparation of Potassium Chloride

A. Write the balanced equation for the reaction between $KHCO_3$ and HCl:

$$KHCO_3 + HCl \longrightarrow KCl + H_2O + CO_2$$

B. Experimental Data and Calculations: Record all measurement to the highest precision of the balance and remember to use the proper number of significant figures in all calculations. (The number 0.004 has only *one* significant figure.)

1. Mass of empty evaporating dish   __66.2671__ g

2. Mass of dish and $KHCO_3$ __68.5176__ g

3. Mass of dish and residue (KCl) after first heating   __68.1386__ g

4. Mass of dish and residue (KCl) after second heating   __68.0678__ g

5. Mass of dish and residue (KCl) after third heating (if necessary) __68.0453__ g

6. Mass of potassium bicarbonate   68.5176   __2.2505__ g
   show calculation set-up   −66.2671

7. Moles of potassium bicarbonate   __0.02248__ mol
   show calculation set-up
$$\text{mol KHCO}_3 = (2.2505 \text{ g}) \left(\frac{1 \text{ mol KHCO}_3}{100.1 \text{ g}} \right) =$$

8. Experimental mass of potassium chloride obtained   __1.7782__ g
   show calculation set-up   68.0453 g − 66.2671 =

9. Experimental moles of potassium chloride obtained   __0.02385__ mol
   show calculation set-up
$$\text{mol KCl} = (1.7782 \text{ g}) \left(\frac{1 \text{ mol KCl}}{74.55 \text{ g}} \right) =$$

10. Theoretical moles of KCl __0.02248__ mol
    show calculation set-up
$$\text{mol KCl} = (2.2505 \text{ KHCO}_3) \left(\frac{1 \text{ mol KHCO}_3}{100.1 \text{ g KHCO}_3} \right) \left(\frac{1 \text{ mol KCl}}{1 \text{ mol KHCO}_3} \right) =$$

11. Theoretical mass of KCl __1.676__ g
    show calculation set-up
$$\text{g KCl} = (0.02248 \text{ mol KCl}) \left(\frac{74.55 \text{ g}}{1 \text{ mol KCl}} \right) =$$

12. Percentage error for experimental mass of KCl vs. theoretical __6.098__ %
    mass of KCl (show calculation set-up)
$$\left(\frac{1.7782 \text{ g} - 1.676 \text{ g}}{1.676 \text{ g}} \right)(100) =$$

QUESTIONS AND PROBLEMS

1. What was done in the experiment to make sure that all the $KHCO_3$ was reacted?

   HCl was added until the fizzing stopped. _or_ An exess of HCl was added to the $KHCO_3$.

2. Why is the mass of KCl recovered less than the starting mass of $KHCO_3$?

   There is an equal number of moles of $KHCO_3$ and KCl (theoretically) but the molar mass of KCl is less than the molar mass of $KHCO_3$.

3. Calculate the moles and grams of HCl present in the 6.0 mL of 6.0 M HCl solution you used.

   $$\text{mol HCl} = (6.0\,\text{mL HCl})\left(\frac{6.0\,\text{mol HCl}}{1000\,\text{mL}}\right) = 0.036$$

   $$\text{mol HCl} = (0.036\,\text{mol HCl})\left(\frac{36.46\,\text{g HCl}}{1\,\text{mol HCl}}\right) = 1.3$$

   _____0.036_____ mol HCl

   _____1.3_____ g HCl

4. Would the 6.0 mL of 6.0 M HCl be sufficient to react with 3.80 g $KHCO_3$? Show supporting calculations and explanation.

   No, 6.0 mL is not enough. It will require 6.3 mL of the acid to react all the $KHCO_3$.

   $$\text{mL HCl} = (3.80\,\text{g KHCO}_3)\left(\frac{1\,\text{mol}}{100.1\,\text{g}}\right)\left(\frac{1\,\text{mol HCl}}{1\,\text{mol KHCO}_3}\right)\left(\frac{1000\,\text{mL HCl}}{6.0\,\text{mol HCl}}\right) = 6.3\,\text{mL HCl}$$

5. Theoretically, why should the moles of $KHCO_3$ and the moles of KCl produced be the same?

   According to the balanced equation, their mole ratio is 1:1 so the moles of KCl produced should be the same.

6. If 3.000 g of K_2CO_3 were used in this experiment (instead of $KHCO_3$),

   (a) What is the balanced equation for the reaction?

   $$K_2CO_3 + 2\,HCl \longrightarrow 2\,KCl + CO_2(g) + H_2O$$

   (b) How many milliliters of 6.0 M HCl would be needed   _____7.2_____ mL HCl

   $$\text{mL HCl} = (3.000\,\text{g K}_2\text{CO}_3)\left(\frac{1\,\text{mol}}{138.2\,\text{g}}\right)\left(\frac{2\,\text{mol HCl}}{1\,\text{mol K}_2\text{CO}_3}\right)\left(\frac{1000\,\text{mL HCl}}{6.0\,\text{mol HCl}}\right)$$

   (c) How many grams of KCl would be formed in the reaction?   _____3.237_____ g KCl

   $$\text{g KCl} = (3.000\,\text{g K}_2\text{CO}_3)\left(\frac{1\,\text{mol}}{138.2\,\text{g}}\right)\left(\frac{2\,\text{mol KCl}}{1\,\text{mol K}_2\text{CO}_3}\right)\left(\frac{74.55\,\text{g KCl}}{1\,\text{mol KCl}}\right)$$

REPORT FOR EXPERIMENT 16

Electromagnetic Energy and Spectroscopy

> ✓ Student results will be variable and will not exactly match the typical student data shown on the key as a guideline for grading.

A. Wave Properties

1. Length of stretched spring _____**360**_____ cm

2. Complete the table below for each of the waves generated by your group.

Form of Wave	Length of Wave Form	No. of cycles	Time (s)	Frequency, cycles/s	Wavelength, cm/wave
(wave form)	$0.5\,\lambda$	50	52	$\dfrac{50}{52} = 0.96$	$2(360) = 720$
(wave form)	$1\,\lambda$	50	24	$\dfrac{50}{24} = 2.1$	360
(wave form)	$1.5\,\lambda$	50	16	$\dfrac{50}{16} = 3.1$	$\dfrac{2}{3}(360) = 240$
(wave form)	$2\,\lambda$	50	13	$\dfrac{50}{13} = 3.8$	$\dfrac{1}{2}(360) = 180$

3. What does the spring have to do with electromagnetic energy?
   The wave pattern of the spring is the same as the path of a photon so the spring can be used to manipulate and measure the properties of a wave.

4. Plot a graph of frequency vs. wavelength for the data produced by the spring. Label the graph with a suitable title, determine an appropriate scale for each axis, and label with units that match the data. See Study Aid 3 for additional help if necessary.

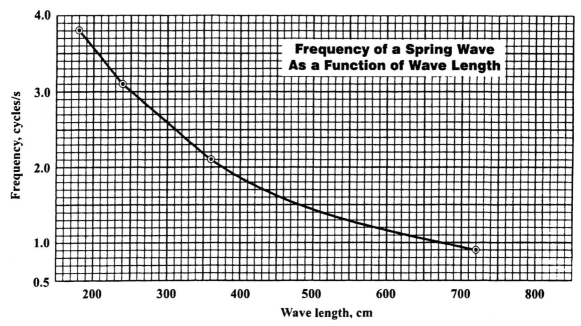

Frequency of a Spring Wave As a Function of Wave Length

(y-axis: Frequency, cycles/s; x-axis: Wave length, cm)

B.   Emission Spectra

1. Use colored pencils and sketch the spectrum observed with the spectroscope for each of the light sources observed.

Light Source	Emission Spectrum Observed
Incandescent Bulb	Continuous spectrum
Fluorescent Bulb	| | | | line spectrum, all colors with space between
Hydrogen Gas	| | | 3 or 4 lines
Neon Gas	| | | | | | | mostly red lines

2. a. Why must the hydrogen vapor lamp be turned on before it gives off light?

   The e⁻ must absorb electical energy to move to a higher energy level so it can fall back to a lower level and emit light.

   b. Why is the spectroscope necessary to observe the hydrogen spectrum?

   It separates the photons into different wavelengths so the spectrum can be seen.

3. Use the colored pencils to color the arrows on the electron transition diagram for H_2 so they correspond to the colors of the visible H_2 spectral lines. Refer to the atomic spectrum chart to match the color to corresponding wavelength then decide the corresponding energy content. Remember that *the length of the arrow is a function of the energy content of the photon released and NOT its wavelength.*

4. Why can we see only 3 or 4 lines in the spectroscope when there are many more arrows in the hydrogen electron transition diagram?

   Photons of wavelengths emitted during e⁻ transitions to $n = 2$ are the only photons of the quanta absorbed by e⁻ of visual pigments in the eye.

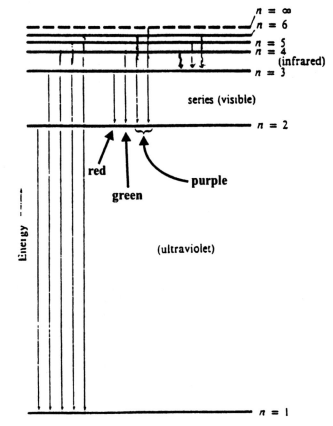

– 136 –

C. Absorption Spectra for colored solutions

1. Record the percent transmittance data measured from the spectrophotometer for each of the solutions shown on the table below.

Percent transmittance

Wavelength, nm	$Ni(NO_3)_2$ (green)	$KMnO_4$ (purple)
350	72.6	0.8
375	37.0	6.0
400	29.2	24.0
425	56.0	34.6
450	84.0	27.2
475	90.6	7.2
500	95.6	1.0
525	94.2	0.4
550	92.6	0.6
575	88.8	10.2
600	79.6	39.0
625	69.6	49.8
650	59.4	58.0
675	58.8	70.4
700	56.4	82.8

2. Graph these data as described in the procedure using either the graph paper provided or a computer if available.

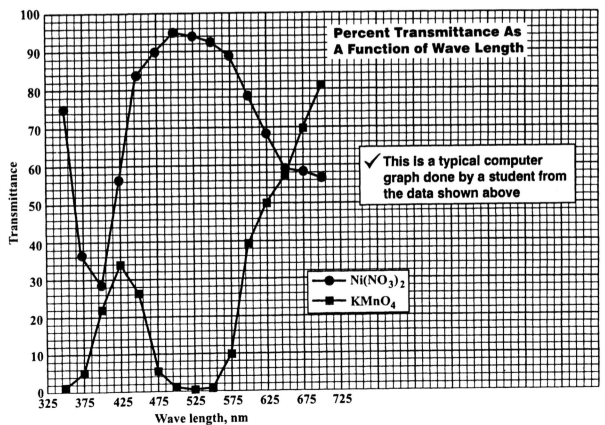

QUESTIONS AND PROBLEMS

1. Draw a diagram on the line below which shows 2.5 transverse waves. Measure the line and calculate the wavelength of a single wave in centimeters.

$$\text{length of line} = \frac{14.0\ \text{cm}}{2.5\ \text{waves}} = 5.6\ \frac{\text{cm}}{\text{wave}}$$

2. If your diagram represents a wave being generated with a spring like the one used in the experiment, and it took 25 seconds to generate 60 of these wave forms, what is the frequency of the wave? Show calculations.

$$\upsilon = \frac{60\ \text{waves}}{25\ \text{s}} = 2.4\ \frac{\text{waves}}{\text{s}}$$

3. What is the difference between an emission spectrum and an absorption spectrum?

   Emission spectrum is the range of photon released by excited e⁻ when they fall from high to low energy levels. Absorption spectrum is the range of photons absorbed when low energy e⁻ jump to higher energy levels.

4. What is the relationship between percent transmittance and absorption?

   % transmittance + % absorption = 100%
   or
   100% − % transmittance = % absorption

5. What is the relationship between percent transmittance and the color of the solutions?

   The wave lengths with the highest percent transmittance correspond to the colors seen for each solution.

6. Where do the photons that are absorbed go when they are absorbed by the solution?

   They are absorbed by e⁻ in the molecules and atoms of the solution.

REPORT FOR EXPERIMENT 17

Lewis Structures and Molecular Models

For each of the following molecules or polyatomic ions, fill out columns A through G using the instructions provided in the procedure section. These instructions are summarized briefly below.

A. Calculate the total number of valence electrons in each formula.

B. Draw a Lewis structure for the molecule or ion which satisfies the rules provided in the procedure.

C. Build a model of the molecule and have it checked by the instructor.

D. Use your model to determine the molecular geometry for this molecule (don't try to guess the geometry without the model): tetrahedral, trigonal pyramidal, trigonal planar, bent, linear

E. Determine the bond angle between the central atom and the atoms bonded to it. If there are only two atoms write "no central atom" in the space provided.

F. Use the electronegativity table to determine the electronegativity of the bonded atoms.
If the bonds are polar, indicate this with a modified arrow ($\longmapsto$) pointing to the more electronegative element.
If the bonds are nonpolar, indicate this with a short line ($-$).
If there are two or more different atoms bonded to the central atom, include each bond.

G. Use your model and your knowledge of the bond polarity to determine if the molecule as a whole is nonpolar or a dipole. If it is polar, write *dipole* in G. If it is not, write *nonpolar*.

	A	B	C	D	E	F	G
Molecule or Polyatomic Ion	No. of Valence Electrons	Lewis Structure		Molecular Geometry	Bond Angles	Bond Polarity	Molecular Dipole or Nonpolar
4 + 4(1) CH_4	8	H | H—C—H | H		tetrahedral	109.5°	C $\longleftarrow$ H	nonpolar
4 + 2(6) CS_2	16	:S̈=C=S̈:		linear	180°	C—S	nonpolar

REPORT FOR EXPERIMENT 17 (continued)   NAME _____ **KEY**

Molecule or Polyatomic Ion	A No. of Valence Electrons	B Lewis Structure	C	D Molecular Geometry	E Bond Angles	F Bond Polarity	G Molecular Dipole or Nonpolar
2(1) + 6 H_2S	8	H–$\ddot{S}$–H		bent	about 105°	H $\leftrightarrow$ S	dipole
2(5) N_2	10	:N≡N:		linear	no central atom	N—N	nonpolar
6 + 6(4) + 2 SO_4^{2-}	32	$[\ddot{O}–S–\ddot{O}]^{2-}$ (with $\ddot{O}$ above and below)		tetrahedral	109.5°	S $\leftrightarrow$ O	nonpolar
3(1) + 6 – 1 H_3O^+	8	$[H–\ddot{O}–H]^+$ with H below		trigonal pyramidal	about 107°	H $\leftrightarrow$ O	dipole
4 + 3(1) + 7 CH_3Cl	14	H–C–$\ddot{Cl}$: with H above and below		tetrahedral	109.5°	C $\leftrightarrow$ Cl, H $\leftrightarrow$ C	dipole
2(4) + 6(1) C_2H_6	14	H–C–C–H with H's		tetrahedral	109.5°	H $\leftrightarrow$ C	nonpolar
2(4) + 4(1) C_2H_4	12	H₂C=CH₂		trigonal planar	120°	H $\leftrightarrow$ C	nonpolar

Molecule or Polyatomic Ion	A No. of Valence Electrons	B Lewis Structure	C	D Molecular Geometry	E Bond Angles	F Bond Polarity	G Molecular Dipole or Nonpolar
2(4) + 2(1) + 2(7) $C_2H_2Cl_2$	24	* Cl–C=C(H)(Cl) structure		trigonal planar	120°	H → C C → Cl	nonpolar (as drawn)
6 + 3(6) + 2 SO_3^{2-}	26	[O–S–O with O below]$^{2-}$		trigonal pyramidal	about 107°	S → O	dipole
4 + 2(1) + 6 CH_2O	12	H–C=O with H		trigonal planar	120°	C → O H → C	dipole
6 + 2(7) OF_2	20	F–O–F		bent	about 105°	O → F	dipole
5 + 2(6) + 1 NO_2^-	18	[O=N–O]$^-$		bent	about 119°	N → O	dipole
2(6) O_2	12	O=O		linear	no central atom	O — O	nonpolar
5 + 3(6) + 1 NO_3^-	24	** O–N=O with O below		trigonal planar	120°	N → O	nonpolar

*More than one possible Lewis structure can be drawn. See questions 1, 2.

**More than one possible Lewis structure can be drawn. See question 3.

QUESTIONS

1. There are three acceptable Lewis structures for $C_2H_2Cl_2$ (*) and you have drawn one of them on the report form. Draw the other two structures and indicate whether each one is nonpolar or a dipole.

(a)

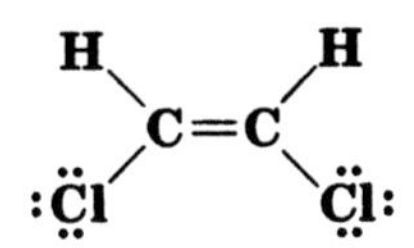

$$H \diagdown \diagup H$$
$$C=C$$
$$:\ddot{C}l \diagup \diagdown \ddot{C}l:$$

dipole

(b)

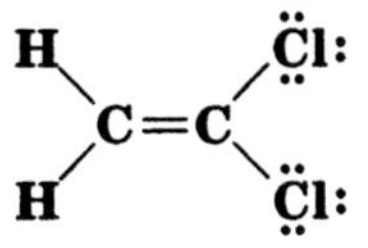

$$H \diagdown \diagup \ddot{C}l:$$
$$C=C$$
$$H \diagup \diagdown \ddot{C}l:$$

dipole

(c)

$$:\ddot{C}l: \diagdown \diagup H$$
$$C=C$$
$$H \diagup \diagdown \ddot{C}l:$$

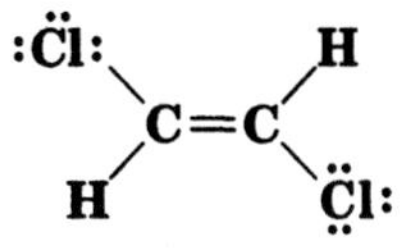

nonpolar

2. Explain why one of the three structures for $C_2H_2Cl_2$ is nonpolar and the other two are molecular dipoles.

The polar C $\leftrightarrow$ Cl bonds cancel each other in 1(c) because the structure is symmetrical. For 1(a) and 1(b) the C $\leftrightarrow$ Cl bonds do not cancel so the molecules do not have a symmetrical charge distribution.

3. There are three Lewis structures for $[NO_3]^-$ (**). Draw the two structures which are not on the report form. Compare the molecular polarity of the three structures.

$$\left[\ddot{O}=N-\ddot{O}: \atop \quad\quad :O: \right]^-$$

$$\left[:\ddot{O}-N=\ddot{O} \atop \quad :O: \right]^-$$

$$\left[:\ddot{O}-N-\ddot{O}: \atop \quad\quad :O: \right]^-$$

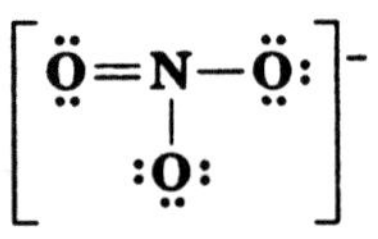

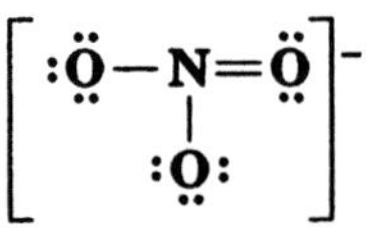

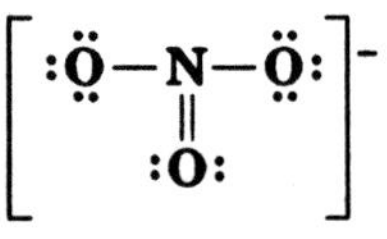

They are all nonpolar because the N $\leftrightarrow$ O bonds cancel each other in these trigonal planar ions. (One of the structures will be on the previous page, the remaning two will be above.)

REPORT FOR EXPERIMENT 18

Boyle's Law

> ✓ Student results will be variable and will not exactly match the typical student data shown on the key as a guideline for grading.

Data Table 1 Show the setup used for every calculated number. Include all units in dimensional analysis setups.

1. Atmospheric pressure   ____747____ mm Hg   ____14.5____ lb/in.^{2}

   $(747 \text{ mm Hg})\left(\dfrac{14.7 \text{ lb/in.}}{760 \text{ mm Hg}}\right)$

2. Inside diameter of syringe   ____2.27____ cm   ____0.894____ in.

   $(2.27 \text{ cm})\left(\dfrac{1 \text{ in.}}{2.54 \text{ cm}}\right) = 0.894 \text{ in.}$

3. Inside radius   ____0.447____ in.

   $\dfrac{0.894 \text{ in.}}{2} = 0.447 \text{ in.}$

4. Area of syringe (cross-section)   ____0.627____ in.^{2}

   $\pi r^2 = (3.14)(0.447 \text{ in.})^2 = 0.627 \text{ in.}^2$

Record measurements and show your calculations for the first applied mass in Data Table 2.

5. First applied mass   ____0.507____ kg

6. Applied mass $(0.507 \text{ kg})\left(\dfrac{2.20 \text{ lb}}{1 \text{ kg}}\right) = 1.10 \text{ lb}$ ____1.10____ lb

7. Applied pressure $\dfrac{1.10 \text{ lb}}{0.627 \text{ in.}^2} = 1.75 \text{ lb/in.}^2$ ____1.75____ lb/in.^{2}

8. Total pressure $14.5 \text{ lb/in.}^2 + 1.75 \text{ lb/in.}^2$ ____16.3____ lb/in.^{2}

9. Volume of air (measured)   ____30.0____ cm^3

10. Pressure-Volume Product, PV $\left(\dfrac{16.3 \text{ lb}}{\text{in.}^2}\right)(30.0 \text{ cm}^3)$ ____489____ lb cm^3/in.^{2}

Data Table 2 Complete the table below for 0.0 applied mass and 4 additional masses. Calculations for the first applied mass should be shown above in 5–10.

Applied Mass, kg	Applied Mass, lb	Applied Pressure, P_{app} (lb/in.^{2})	Total Pressure $P_{gas} = P_{atm} + P_{app}$ (lb/in.^{2})	Volume of Air, V (cm^3)	Pressure-Volume Product, PV (lb cm^3/in.^{2})
0	0	0	14.5	32.5	471
0.507	1.10	1.75	16.3	30.0	489
2.31	5.08	8.10	22.6	22.0	497
2.82	6.20	9.89	24.4	20.0	488
3.83	8.43	13.4	27.9	17.8	496

Average PV __488 lb cm^3/in.^{2}__

QUESTIONS AND PROBLEMS

1. What is the independent variable in this experiment? Explain your choice.

   Pressure. It is being controlled by the experimenter.

2. Why must the temperature be constant during this experiment?

   Temperature will also cause changes in volume. (If temperature is not held constant it is difficult to tell if volume changes are due to pressure or temp.)

3. What part of the tabulations in the data tables proves Boyle's Law? How?

   PV products should be close to the average value

4. Plot Total Pressure vs. Volume of Air for the five values in Data Table 2. Use the graph paper provided or attach a computer-generated graph to your report form. If necessary, refer to Study Aid 3 for instructions on how to complete either type of graph.

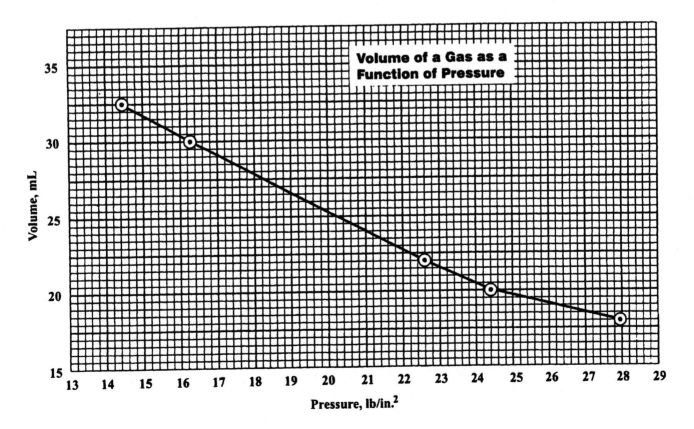

5. What is the relationship between gas pressure and volume shown by the data in your graph?

   Inverse relationship. As pressure increases, volume decreases.

6. If you repeated this experiment at a lower temperature (for example in a walk-in refrigerator), how would the P vs. V curve obtained differ from the curve on your graph?

   The PV curve would be parallel to and below the curve on the graph (because volume would be less at each pressure).

7. Given the following data: volume of air without any applied pressure, 25.0 cm³; inside diameter of the syringe, 3.20 cm; barometric pressure, 630 mmHg. Show setups and answers for each of the following problems based on this data.

   a. What is the cross-sectional area of the syringe in in.²?

   $$\text{diameter} = (3.20\,\text{cm})\left(\frac{1\,\text{in.}}{2.54\,\text{cm}}\right) = 1.26\,\text{in.}$$

   $$A = \pi r^2 = (3.14)\left(\frac{1.26\,\text{in.}}{2}\right)^2 = 1.25\,\text{in.}^2$$

   b. What is the barometric pressure in lb/in.²?

   $$(630.\,\text{mm Hg})\left(\frac{14.7\frac{\text{lb}}{\text{in.}^2}}{760.\,\text{mm Hg}}\right) = 12.2\,\frac{\text{lb}}{\text{in.}^2}$$

   c. A brick with a mass of 6.00 lb is placed on the barrel of the syringe. What is the total pressure, P_{gas}, of the gas in the syringe after adding the brick?

   $$P_{app} = \frac{\text{mass}}{A} = \frac{6.00\,\text{lb}}{1.25\,\text{in.}^2} = 4.8\,\frac{\text{lb}}{\text{in.}^2}$$

   $$P_{gas} = P_{app} + P_{atm} = 4.8\,\frac{\text{lb}}{\text{in.}^2} + 12.2\,\frac{\text{lb}}{\text{in.}^2} = 17.0\,\frac{\text{lb}}{\text{in.}^2}$$

   d. What is the change in volume after adding the brick?

   $$P_1V_1 = P_2V_2 \qquad\qquad V_2 = \frac{P_1V_1}{P_2}$$

   $$V_2 = \frac{(12.2\,\text{lb/in.}^2)(25.0\,\text{cm}^3)}{17.0\,\text{lb/in.}^2} = 17.9\,\text{cm}^3$$

   $$\text{Change in volume} = V_2 - V_1 = 17.9\,\text{cm}^3 - 25.0\,\text{cm}^3 = -7.1\,\text{cm}^3$$

NAME _____ **KEY** _____

SECTION _____ DATE _____

REPORT FOR EXPERIMENT 19

INSTRUCTOR _____

Charles' Law

> ✓ Student results will be variable and will not exactly match the typical student data shown on the key as a guideline for grading.

Data Table

	Trial 1	Trial 2
Temperature of boiling water, T_H	__95.0__ °C, __368__ K	_____ °C, _____ K
Temperature of cold water, T_L	__14.0__ °C, __287__ K	_____ °C, _____ K
Volume of water collected in flask (decrease in the volume of air due to cooling)	31.1 mL	
Volume of air at higher temperature, V_H (volume of flask measured only after Trial 2)	149 mL	
Volume of wet air at lower temperature (volume of flask less volume of water collected), V_{exp}	149 mL − 31.1 mL 118 mL	
Atmosphere pressure, P_{atm} (barometer reading)	736.1 torr	
Vapor pressure of water at lower temperature, P_{H_2O} (see Appendix 6)	12.1 torr	

REPORT FOR EXPERIMENT 19 (continued)   NAME _____ KEY _____

CALCULATIONS: In the spaces below, show calculation setups for Trial 1 only. Show answers for both trials in the boxes

	Trial 1	Trial 2

1. Corrected experimental volume of dry air at the lower temperature calculated from data obtained at the lower temperature.

   (a) Pressure of dry air (P_{DA})

$$P_{DA} = P_{Atm} - P_{H_2O}$$

724.0 torr

736.1 torr − 12.1 torr = 724.0 torr

   (b) Corrected experimental volume of dry air (lower temperature).

$$V_{DA} = (V_{exp})\left(\frac{P_{DA}}{P_{Atm}}\right) =$$

116 mL

$$V_{DA} = (118\,mL)\left(\frac{724.0\,torr}{736.1\,torr}\right) = 116\,mL$$

2. Predicted volume of dry air at lower temperature V_L calculated by Charles' law from volume at higher temperature (V_H).

$$V_L = (V_H)\left(\frac{T_L}{T_H}\right)$$

116 mL

$$V_L = (149\,mL)\left(\frac{287\,K}{368\,K}\right) = 116\,mL$$

3. Percentage error in verification of Charles' law.

$$\%\ error = \left(\frac{V_{DA} - V_L}{V_L}\right)(100) =$$

0%

$$= \left(\frac{116 - 116}{116}\right)(100) = 0\%$$

4. Comparison of experimental V/T ratios. (Use dry volumes and absolute temperatures.)

   (a) $\dfrac{V_H}{T_H} = \dfrac{149\,mL}{368\,K} = 0.405\,mL/K$

0.405 mL/K

   (b) $\dfrac{V_{DA}}{T_L} = \dfrac{116\,mL}{287\,K} = 0.405\,mL/K$

0.405 mL/K

− 166 −

5. On the graph paper provided, plot the volume-temperature values used in Calculation 4. Temperature data **must be in °C**. Draw a straight line between the two plotted points and extrapolate (extend) the line so that it crosses the temperature axis.

QUESTIONS AND PROBLEMS

1. (a) In the experiment, why are the water levels inside and outside the flask equalized before removing the flask from the cold water?

    To equalize the pressure inside the flask with that of the atmosphere.

   (b) When the water level is higher inside than outside the flask, is the gas pressure in the flask higher than, lower than, or the same as, the atmospheric pressure? (specify which)

    _____ **lower than** _____

2. A 125 mL sample of dry air at 230°C is cooled to 100°C at constant pressure. What volume will the dry air occupy at 100°C?

   $$V_2 = (125\,mL)\left(\frac{373\,K}{503\,K}\right) = 92.7\,mL$$

    _____ **92.7** _____ mL

3. A 250 mL container of a gas is at 150°C. At what temperature will the gas occupy a volume of 125 mL, the pressure remaining constant?

   $$T_2 = (423\,K)\left(\frac{125\,mL}{250.\,mL}\right) = 212\,K\;\underline{or}\;-61°C$$

    _____ **212 K or −61°C** _____ °C

4. (a) An open flask of air is cooled. Answer the following:

   1. Under which conditions, before or after cooling, does the flask contain more gas molecules?

    _____ **after** _____

   2. Is the pressure in the flask at the lower temperature the same as, greater than, or less than the pressure in the flask before it was cooled?

    _____ **same** _____

(b) An open flask of air is heated, stoppered in the heated condition, and then allowed to cool back to room temperature. Answer the following:

1. Does the flask contain the same, more, or fewer gas molecules now compared to before it was heated?

_____ **less** _____

2. Is the volume occupied by the gas in the flask approximately the same, greater, or less than before it was heated?

_____ **same** _____

3. Is the pressure in the flask the same, greater, or less than before the flask was heated?

_____ **less** _____

4. Do any of the above conditions explain why water rushed into the flask at the lower temperature in the experiment? Amplify your answer.

Yes. The pressure inside the flask was less than atmospheric pressure so when the flask was opened, water rushed in.

5. On the graph you plotted,

(a) At what temperature does the extrapolated line intersect the x-axis?

about −275 °C

(b) At what temperature does Charles' law predict that the extrapolated line should intersect the x-axis?

_____ **−273** _____ °C

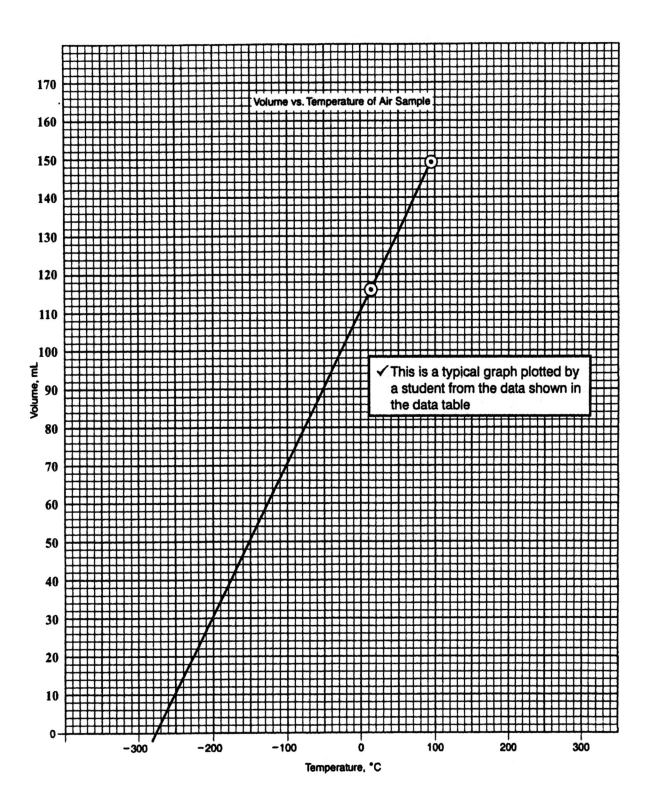

Volume vs. Temperature of Air Sample

✓ This is a typical graph plotted by a student from the data shown in the data table

Temperature, °C

REPORT FOR EXPERIMENT 20

Liquids: Vapor Pressure and Boiling Points

> ✓ Student results will be variable and will not exactly match the typical student data shown on the key as a guideline for grading.

A. Cooling Effect of Evaporation

Room Temperature ___**21.5°C**___

Data Table A.

Liquid wetting thermometer bulb	Lowest Temperature observed, °C	Normal Boiling Point (See Appendix 7)
Acetone	−3.0°C	56.2
Methanol	1.0°C	64.6
Ethanol	8.5°C	78.5
Water	12.0°C	100.

1. Why does the temperature drop when the thermometer covered with wet filter paper is suspended in the air?

   Evaporation consumes heat. Thus evaporation of liquids on the filter paper cools the paper and the thermometer <u>or</u> the "hotter" molecules of liquid evaporate first leaving the "cooler" liquid molecules behind which lowers the temperature of the thermometer bulb.

2. On the basis of the observed behavior of these four liquids, what is the relationship between the normal boiling points and effectiveness of liquids in cooling by evaporation?

   The lower the normal boiling point, the more effective the liquid will be in cooling by evaporation.

B. Relationship of Boiling Point and Vapor Pressure

Data Table B. Circle the temperature at which boiling stopped in each tube.

Temperature Readings, °C			
Closed Tube	Open Tube	Closed Tube	Open Tube
105	105	70	70
100	(100)	65	63
92	95		60
90	91		55
(85)	86		50
80	79		43
75	75		

1. How much water was in the delivery tube (length of water column above the screw clamp)?

   (a) at the time of closing the screw clamp   ____**0**____ cm

   (b) after boiling has stopped   __**variable**__ cm

2. When the water had cooled to 85°C in the open tube and you touched the upper part of both test tubes, which was hotter?   __**Closed tube**__
   Why?
   Water in the closed tube was still boiling, giving off vapor. Vapor condensing on the walls of the tube liberates heat.

3. During the cooling, what happened to the rubber tubing at the top of the closed tube? Why?
   It collapsed because the pressure inside the system was less than atmospheric pressure.

4. When you opened the screw clamp under water, what happened? Why?
   Water rushed in because the lower pressure inside the system could not oppose the atmospheric pressure pushing the water in.

5. Why did the water in the closed test tube continue to boil at lower temperatures than the water in the open test tube?
   Vapor pressure of the hot water exeeds (or is equal to) the pressure of the atmosphere inside the closed test tube.

6. Why did the water in the closed test tube cool faster than the water in the open test tube?
   Because the rate of evaporation was greater in the closed tube where the water was still boiling and the evaporation of a liquid consumes heat.

7. Explain clearly what caused the reduced pressure inside the closed test tube during cooling.
   As the temperature drops, steam (water vapor) condenses to a liquid, reducing the amount of vapor present and thus decreasing the vapor pressure in the closed tube.

8. Why was there more water in the delivery tube after the apparatus cooled (before the clamp was opened) than at the time it was clamped shut.
   Steam condensed to liquid water in the delivery tube.

C.   Effect of Vapor Pressure Change and Atmospheric Pressure

1. (a) What did you observe at the mouth of the can just before it was stoppered?

    Condensing steam (cloud) <u>or</u> condensing water vapor.

   (b) What does this indicate about the composition of the gas in the can when the can was stoppered?

    The gas in the can was mostly steam.

2. What happened to the can after heating was stopped and it was stoppered tightly?

   The can collapsed.

3. If the can had been stoppered and heating stopped when the water first began to boil, how might the results have been different? Why?

   The can might not have collapsed as much because a considerable amount of air would still have been in the can.

4. If the can had been allowed to cool about three minutes before stoppering, how might the results have been different? Why?

   The can might not have collapsed as much. Air would have entered the unstoppered can while it was cooling before it was stoppered.

5. If the can had been stoppered about three minutes before heating was stopped, how might the results have been different? Why?

   The can might have bulged or exploded or the stopper might have been blown out. Since the steam could not escape, the pressure inside the can would continue to build up beyond the atmospheric pressure.

REPORT FOR EXPERIMENT 21

Molar Volume of a Gas

> ✓ Student results will be variable and will not exactly match the typical student data shown on the key as a guideline for grading.

Measurements

1. Concentration of H_2O_2 _____ 3.0 _____ %

2. Volume of H_2O_2 added to the generator _____ 3.0 _____ mL

3. Barometric pressure _____ 744.2 _____ mm Hg

4. Water temperature _____ 21.0 _____ °C _____ 294 _____ K

5. Gas volume collected in the graduated cylinder _____ 35.5 _____ mL

Calculations: *Show the setup for every calculation.*

1. Oxygen volume generated by the reaction _____ 32.5 _____ mL

   35.5 mL − 3.0 mL = 32.5 mL

2. Moles of oxygen generated by the decomposition of H_2O_2 _____ 0.0013 _____ mol

   (a) balanced equation for the reaction $\quad 2\,H_2O_2 \xrightarrow{\text{MnO}_2} 2\,H_2O + O_2$

   (b) stoichiometric setup using continuous calculation method

   $$\textbf{mol } O_2 = (3.0\text{ mL }H_2O_2(aq))\left(\frac{3.0\text{ g }H_2O_2}{100\text{ mL }H_2O_2(aq)}\right)\left(\frac{1\text{ mol }H_2O_2}{34.02\text{ g }H_2O_2}\right)\left(\frac{1\text{ mol }O_2}{2\text{ mol }H_2O_2}\right) = 0.0013\text{ mol}$$

3. Pressure of dry gas in the graduated cylinder

   (a) vapor pressure of water @ 21.0°C _____ 18.6 _____ mm Hg

   (b) pressure of dry gas $\quad$ **744.2 mm Hg − 18.6 mm Hg =** _____ 725.6 _____ mm Hg

4. Experimental volume of dry oxygen converted to STP _____ 28.8 _____ mL

   $$\frac{V_1P_1}{T_1} = \frac{V_2P_2}{T_2} \qquad \frac{(32.5\text{ mL})(725.6\text{ mm Hg})}{294\text{ K}} = \frac{(V_2)(760\text{ mm Hg})}{273\text{ K}} \qquad V_2 = 28.8\text{ mL}$$

5. Experimental molar volume _____ 22 _____ L/mol

   $$\frac{L}{mol} = \left(\frac{28.8\text{ mL}}{0.0013\text{ mol}}\right)\left(\frac{1\text{ L}}{1000\text{ mL}}\right) = 22$$

6. Theoretical molar volume (L/mol) _____ 22.4 _____ L/mol

7. Percent error $\left(\dfrac{22.4\text{ L/mol} - 22\text{ L/mol}}{22.4\text{ L/mol}}\right)(22.4\text{ L/mol}) =$ _____ 1.8 _____ %

QUESTIONS AND PROBLEMS

1. The curved surface of an aqueous solution is called _____ **meniscus** _____.
   If the top of this surface were used to measure the volume of the gas rather than the bottom of the surface, the volume of the gas above the liquid be _____ **too small** _____
   (too large or too small)

2. Why was it necessary to subtract the volume of the H_2O_2 solution injected into the generator from the volume of gas collected in the graduated cylinder?
   This volume of water displaced was not due to the decomposition reaction on which moles of gas generated is calculated so it must be substracted from gas volume collected.

3. If a student did this experiment using 5.0 mL of 10.% hydrogen peroxide, how many moles of O_2 would be generated?

$$\textbf{mol } O_2 = (5.0\,\textbf{mL } H_2O_2(aq))\left(\frac{10.\,\textbf{g } H_2O_2}{100.\,\textbf{mL } H_2O_2(aq)}\right)\left(\frac{1\,\textbf{mol } H_2O_2}{34.02\,\textbf{g}}\right)\left(\frac{1\,\textbf{mol } O_2}{2\,\textbf{mol } H_2O_2}\right) = \underline{0.0073}\,\textbf{mol } O_2$$

4. Why was it not necessary to measure the MnO_2 precisely when we have to be so careful about measuring the volume of H_2O_2 and the volume of oxgyen collected.

   MnO_2 is a catalyst, not a product or a reactant. (It speeds up the reaction but remains unchanged.)

5. At 20.0°C, a student collects H_2 gas in a gas collecting tube. The barometric pressure is 755.2 mm Hg and the water levels inside and outside the tube are exactly equal.

   (a) What is the total gas pressure in the gas collecting tube?

   _____**755.2**_____ mm Hg

   (b) What is the pressure of the water vapor in the gas collecting tube?

   _____**17.5**_____ mm Hg

   (c) What is the pressure of the dry hydrogen in the gas collecting tube?

   755.2 mm Hg − 17.5 mm Hg = 737.7 mm Hg   _____**737.7**_____ mm Hg

6. What would be the effect on the molar volume if hydrogen instead of oxygen gas had been collected during this experiment? Explain your answer.

   Molar volume would be the same because equal volumes of gases at the same conditions of T and P contain equal numbers of molecules or moles.

7. After correction to standard conditions, the volume of a gas collected was 43.8 mL. If this volume represents 0.00184 mol, what is the percent error for the experimental molar volume?

$$\frac{\textbf{L}}{\textbf{mol}} = \left(\frac{43.8\,\textbf{mL}}{0.00184\,\textbf{mol}}\right)\left(\frac{1\,\textbf{L}}{1000\,\textbf{mL}}\right) = \textbf{23.8 L/mol}$$

$$\left(\frac{23.8\,\textbf{L/mol} - 22.4\,\textbf{L/mol}}{22.4\,\textbf{L/mol}}\right)(100) = \textbf{6.25\% error}$$

REPORT FOR EXPERIMENT 22

Neutralization–Titration I

> ✓ Student results will be variable and will not exactly match the typical student data shown on the key as a guideline for grading.

Data Table

	Sample 1	Sample 2	Sample 3 (if needed)
Mass of flask and KHP	94.7994 g	95.3800 g	
Mass of empty flask	93.6616 g	94.2144 g	
Mass of KHP	1.1378 g	1.1656 g	
Final buret reading	18.25 mL	36.94 mL	
Initial buret reading	0.15 mL	18.24 mL	
Volume of base used	18.10 mL	18.70 mL	

CALCULATIONS: In the spaces below show calculation setups for Sample 1 only. Show answers for both samples in the boxes. Remember to use the proper number of significant figures in all calculations. (The number 0.005 has only one significant figure.)

	Sample 1	Sample 2	Sample 3 (if needed)
1. Moles of acid (KHP, Molar mass = 204.2) $(1.1378 \text{ g KHP})\left(\dfrac{1 \text{ mol}}{204.2 \text{ g}}\right) =$	0.005572 mol KHP	0.005708 mol KHP	
2. Moles of base used to neutralize (react with) the above number of moles of acid $(0.005572 \text{ mol KHP})\left(\dfrac{1 \text{ mol NaOH}}{1 \text{ mol KHP}}\right) =$	0.005572 mol NaOH	0.005708 mol NaOH	
3. Molarity of base (NaOH) $\dfrac{0.005572 \text{ mol NaOH}}{0.01810 \text{ L}} =$	0.3078 mol/L	0.3052 mol/L	

4. Average molarity of base _____**0.3065 M**_____

5. Unknown base number _____**A**_____

QUESTIONS AND PROBLEMS

1. If you had added 50 mL of water to a sample of KHP instead of 30 mL, would the titration of that sample then have required more, less, or the same amount of base? Explain.

   Same amount of base. Dilution would not have changed the number of moles of KHP, so the titration would require the same amount of NaOH.

2. A student weighed out 1.106 g of KHP How many moles was that?

   $$(1.106\ g\ KHP)\left(\frac{1\ mol}{204.2\ g}\right) =$$

   _____0.005416 <u>or</u> 5.416 $\times$ 10^{-3}_____ mol

3. A titration required 18.38 mL of 0.1574 M NaOH solution. How many moles of NaOH were in this volume?

   $$(0.01838\ L)\left(\frac{0.1574\ mol}{L}\right) =$$

   _____0.002893 or 2.893 $\times$ 10^{-3}_____ mol

4. A student weighed a sample of KHP and found it weighed 1.276 g. Titration of this KHP required 19.84 mL of base (NaOH). Calculate the molarity of the base.

   $$(1.276\ g\ KHP)\left(\frac{1\ mol}{204.2\ g}\right)\left(\frac{1\ mol\ NaOH}{1\ mol\ KHP}\right)\left(\frac{1}{0.01984\ L}\right) = 0.3150\ mol/L$$

   _____0.3150_____ M

5. Forgetful Freddy weighed his KHP sample, but forgot to bring his report sheet along, so he recorded the mass of KHP on a paper towel. During his titration, which required 18.46 mL of base, he spilled some base on his hands. He remembered to wash his hands, but forgot about the data on the towel, and used it to dry his hands. When he went to calculate the molarity of his base, Freddy discovered that he didn't have the mass of his KHP. His kindhearted instructor told Freddy that his base was 0.2987 M. Calculate the mass of Freddy's KHP sample.

   $$(0.01846\ L\ NaOH)\left(\frac{0.2987\ mol}{L}\right)\left(\frac{1\ mol\ KHP}{1\ mol\ NaOH}\right)\left(\frac{204.2\ g}{mol}\right) = 1.126\ g\ KHP$$

   _____1.126_____ g

6. What mass of solid NaOH would be needed to make 645 mL of Freddy's NaOH solution?

   $$(0.645\ L)\left(\frac{0.2987\ mol}{L}\right)\left(\frac{40.00\ g\ NaOH}{mol}\right) = 7.71\ g\ NaOH$$

   _____7.71_____ g

NAME **KEY**

SECTION _____ DATE _____

INSTRUCTOR _____

REPORT FOR EXPERIMENT 23

Neutralization–Titration II

> ✓ Student results will be variable and will not exactly match the typical student data shown on the key as a guideline for grading.

A. Molarity of an Unknown Acid

Data Table

	Sample 1 mL		Sample 2 mL		Sample 3 (if needed)	
	Acid*	Base	Acid*	Base	Acid*	Base
Final buret reading		19.84		39.59		
Initial buret reading		0.20		19.84		
Volume used	10.00	19.64	10.00	19.75		

*If a pipet is used to measure the volume of acid, record only in the space for volume used.

Molarity of base (NaOH) **0.3065 M (from Exp. 22)**

CALCULATIONS: In the spaces below, show calculation setups for Sample 1 only. Show answers for both samples in the boxes.

	Sample 1	Sample 2	Sample 3 (if needed)

1. Moles of base (NaOH) (if needed)

$$(0.01964\,L)\left(\frac{0.3065\,mol}{1\,L}\right) =$$

| 0.006020 mol | 0.006053 mol | |

2. Moles of acid used to neutralize (react with) the above number of moles of base

$$(0.006020\,mol\,NaOH)\left(\frac{1\,mol\,HCl}{1\,mol\,NaOH}\right) =$$

| 0.006020 mol | 0.006053 mol | |

3. Molarity of acid

$$\frac{0.006020\,mol}{0.01000\,L} =$$

| 0.6020 M | 0.6053 M | |

4. Average molarity of acid **0.6037 M**

5. Unknown acid number **A**

– 199 –

B. **Acetic Acid Content of Vinegar**

Data Table

	Sample 1 mL		Sample 2 mL		Sample 3 (if needed)	
	Vinegar*	Base	Vinegar*	Base	Vinegar*	Base
Final buret reading		27.30		35.49		
Initial buret reading		0.15		8.25		
Volume used	10.00	27.15	10.00	27.24		

*If a pipet is used to measure the volume of vinegar, record only in the space for volume used.

Molarity of base (NaOH) _____ **0.3065 M** _____   Vinegar number _____

CALCULATIONS: In the spaces below, show calculation setups for Sample 1 only. Show answers for both samples in the boxes.

	Sample 1	Sample 2	Sample 3 (if needed)

1. Moles of base (NaOH)

$$(0.02715\,\text{L})\left(\frac{0.3065\,\text{Mol}}{1\,\text{L}}\right) =$$

0.008321 mol	0.008349 mol	

2. Moles of acid ($HC_2H_3O_2$) used to neutralize (react with) the above number of moles of base

$$(0.008321\,\text{mol NaOH})\left(\frac{1\,\text{mol}\,HC_2H_3O_2}{1\,\text{mol NaOH}}\right)$$

0.008321 mol	0.008349 mol	

3. Molarity of acetic acid in the vinegar

$$\frac{0.008321\,\text{mol}}{0.01000\,\text{L}} =$$

0.8321 M	0.8349 M	

4. Average molarity of acetic acid in the vinegar _____ **0.8335 M** _____

5. Grams of acetic acid per liter (from average molarity) _____ **50.05 g/L** _____

$$\left(\frac{0.8335\,\text{mol}}{\text{L}}\right)\left(\frac{60.05\,\text{g}\,HC_2H_3O_2}{\text{mol}}\right) =$$

6. Mass percent acetic acid in vinegar sample (density of vinegar = 1.005 g/mL) _____ **4.980%** _____

$$\left(\frac{50.05\,\text{g}}{1000.\,\text{mL}}\right)\left(\frac{1\,\text{mL}}{1.005\,\text{g}}\right) = 004980 \times 100 = 4.980\%$$

NAME **KEY**

SECTION _____ DATE _____

REPORT FOR EXPERIMENT 24

INSTRUCTOR _____

Chemical Equilibrium–Reversible Reactions

Refer to equilibrium equations in the discussion when answering these questions.

A. Saturated Sodium Chloride

1. What is the evidence for a shift in equilibrium?

    Formation of salt crystals <u>or</u> precipitate

2. Which ion caused the equilibrium to shift?   **chloride <u>or</u> Cl⁻**

3. In which direction did the equilibrium shift?   **left**

4. If solid sodium hydroxide were added to neutralize the hydrochloric acid, would this reverse the reaction and cause the precipitated sodium chloride to redissolve? Explain.

    No. The neutralization only involves the H⁺ and OH⁻ ions and does not affect the equilibrium. (In addition, the increased Na⁺ (from NaOH) favors precipitation of solid NaCl.)

B. Saturated Ammonium Chloride

1. What is the evidence for a shift in equilibrium?

    Formation of salt crystals <u>or</u> precipitate

2. In which direction did the equilibrium shift?   **left**

3. Which ion caused the equilibrium to shift?   **chloride <u>or</u> Cl⁻**

C. Iron(III) Chloride plus Potassium Thiocyanate

1. What is the evidence for a shift in equilibrium when iron(III) chloride is added to the stock solution?

    Red color becomes more intense <u>or</u> darker

2. What is the evidence for a shift in equilibrium when potassium thiocyanate is added to the stock solution?

    Red color becomes more intense <u>or</u> darker

3. (a) What is the evidence for a shift in equilibrium when silver nitrate is added to the stock solution? (The formation of a precipitate is not the evidence since the precipitate is not one of the substances in the equilibrium.)

Red color became lighter

(b) The change in concentration of which ion in the equilibrium caused this equilibrium shift?

_____ **SCN⁻** _____

(c) Write a net ionic equation to illustrate how this concentration change occurred.

$Ag^+ + SCN^- \longrightarrow AgSCN(s)$ **(SCN⁻ removed as a ppt.)**

(d) When the mixture in C.4 was divided and further tested, what evidence showed that the mixture still contained Fe^{3+} ions in solution?

The red color appeared when KSCN was added.

D. Copper(II) Sulfate Solution with Ammonia

1. What was the evidence for the first shift in equilibrium when the $NH_3(aq)$ was added dropwise to the Cu^{2+} solution?

The clear blue solution became darker blue and cloudy

2. (a) Explain how adding more $NH_3(aq)$ caused the equilibria to shift again.

The addition of more $NH_3(aq)$ dissolved the insoluble $Cu(OH)_2$ by forming a soluble amine complex ion $[Cu(NH_3)_4^{2+}]$.

(b) What did you observe in the Cu^{2+} system to indicate that the shift had occured?

The solution changed from a blue cloudy appearance to a clear deeper blue color.

3. (a) Explain how 3 M sulfuric acid caused the equilibria to shift back again?

The H^+ from the acid reacted with the NH_3 forming NH_4^+, causing the equilibrium to shift to the left.

(b) What did you observe to indicate that the reaction shifted to the left?

The deep blue color changed back to the pale blue color of $Cu(H_2O)_4^{2+}$.

E. **Cobalt(II) Chloride Solution**

1. What was the evidence for a shift in equilibrium when conc. hydrochloric acid was added to the cobalt chloride solution?

    The solution gradually changed from pink to an intense blue color.

2. (a) Write the equilibrium equation for this system.

    $$Co(H_2O)_6^{2+} + 4\,Cl^- \rightleftharpoons CoCl_4^{2-} + 6\,H_2O$$

    (b) State whether the concentration of each of the following substances was increased, decreased, or unaffected when the conc. hydrochloric acid was added to cobalt chloride solution.

    $Co(H_2O)_6^{2+}$ ___**decreased**___, Cl^- ___**increased**___, $CoCl_4^{2-}$ ___**increased**___

3. (a) What did you observe when ammonium chloride was added to cobalt chloride solution?

    The solution became less pink and appeared to have some bluish or purplish color in it. All the NH$_4$Cl did not dissolve.

    (b) What did you observe when this mixture was heated?

    The color changed to an intense blue. (Some or all of the NH$_4$Cl dissolved.)

    (c) Explain why heating the mixture caused the equilibrium to shift.

    More NH$_4$Cl dissolved in the heated solution resulting in an increased Cl$^-$ ion concentration and shifting the equilibrium to the right. (indicated by the intense blue color formed)

    (d) What did you observe when the mixture was cooled?

    Salt (NH$_4$Cl) crystallized out of solution and the color changed from blue towards pink.

    (e) Explain why cooling the mixture caused the equilibrium to shift.

    Crystallization of NH$_4$Cl removed Cl$^-$ ions from the solution causing the equilibrium to shift (to the left)

F. Ammonia Solution

1. What is the evidence for a shift in equilibrium when ammonium chloride was added to the stock solution?

    The pink color disappeared.

2. Explain, in terms of the equilibrium, the results observed when hydrochloric acid was added to the stock solution.

    The H^+ ions from the acid reduced the concentration of OH^- ions (and NH_3) in the equilibrium until the solution was no longer alkaline to phenolphthalein.

3. State whether the concentration of each of the following was increased, decreased, or was unaffected when dilute hydrochloric acid was added to the ammonia stock solution:

    NH_3 **decreased** , NH_4^+ **increased** , OH^- **decreased** , Pink color **decreased**

4. (a) In which direction would the equilibrium shift if sodium hydroxide were added to the ammonia stock solution?

    The equilibrium would shift to the left.

    (b) Would the sodium hydroxide tend to decrease the color intensity? Explain.

    No. NaOH causes the OH^- ion concentration to increase and makes the solution more basic.

5. Would boiling the ammonia solution have any effect on the equilibrium? Explain.

    Yes. NH_3 is volatile and would escape from the solution when heated, thereby shifting the equilibrium to the left.

NAME **KEY**

SECTION _____ DATE _____

INSTRUCTOR _____

REPORT FOR EXPERIMENT 25

Heat of Reaction

> ✓ Student results will be variable and will not exactly match the typical student data shown on the key as a guideline for grading.

A. Hydration of Concentrated Sulfuric Acid

1. Temperature of water.   19.6°C

   Temperature of water after adding acid.   43.0°C

   Temperature change.   +23.4°C

2. Is this an endothermic or exothermic reaction?   exothermic

3. In this reaction, what is the major change, in terms of bonds made or broken, that causes the heat effect?

   Formation of bonds between the sulfuris acid and the water <u>or</u> the formation of chemical bonds

4. If you had used dilute sulfuric acid rather than concentrated sulfuric acid, would you expect the temperature change to be greater or less?

   Why?   less

   In the diluite sulfuric acid, most of the acid ions are already hydrated.

B. Dissolving of Ammonium Chloride

1. Temperature of water.   19.2°C

   Temperature of water after adding salt.   9.1°C

   Temperature change.   −10.1°C

2. Is this an endothermic or exothermic reaction?   endothermic

3. In this reaction, what is the major change, in terms of bonds made or broken, that causes the heat effect?

   The energy needed to break the ionic bonds of the NH_4Cl lattice structure is taken from the water, which results in a lower water temperature.

4. What other heat effect is present in this reaction, acting in the opposite direction?

   The hydration energy <u>or</u> the formation of bonds between the ions and water.

5. If you had used 40 mL of water and 6 g of ammonium chloride, rather than the 20 mL and 3 g in the experiment, would you expect to get a larger, smaller, or identical temperature change?

   Why?   **nearly identical**

   In both cases the H_2O and NH_4Cl are in the same ratio (20 : 3)

C.   Neutralization Reactions

Data Table

		Temp. Before Adding NaOH	Temp. After Adding NaOH	Temp. Change
1. 20.0 mL H_2O	Trial 1	19.0	23.4	4.4
10.00 mL HCl	Trial 2	18.9	23.0	4.1
10.0 mL NaOH	Average	19.0	23.3	(4.3)
2. 20.0 mL H_2O	Trial 1	18.8	23.0	4.2
10.0 mL HNO_3	Trial 2	19.0	23.1	4.1
10.0 mL NaOH	Average	18.9	23.1	(4.2)
3. 20.0 mL H_2O	Trial 1	19.0	23.0	4.0
10.0 mL H_2SO_4	Trial 2	19.0	22.9	3.9
10.0 mL NaOH	Average	19.0	23.0	(4.0)
4. 10.0 mL H_2O	Trial 1	19.2	23.0	3.8
20.0 mL HCl	Trial 2	19.1	23.2	4.1
10.0 mL NaOH	Average	19.2	23.1	(3.9)
5. 10.0 mL H_2O	Trial 1	18.8	26.4	7.6
10.0 mL HCl	Trial 2	19.0	27.0	8.0
20 mL NaOH	Average	18.9	26.7	(7.8)

QUESTIONS AND PROBLEMS

1. (a) What was the average temperature change for neutralization reactions #1, 2, and 3?

$$4.3 + 4.2 + 4.0 = \frac{12.5}{3}$$ ___4.2°C___

   (b) What do you conclude from this?

   Since they are all similar, the reaction is not dependent on which of the three acids was used <u>or</u> same number of bonds are formed regardless of which acid was used.

2. How many moles of $H_2O(l)$ were formed in each neutralization reaction (#1, 2, and 3)?
   Show calculation set-ups:

   $$mol\ H_2O = (10.0\ mL\ NaOH)\left(\frac{1.25\ mol}{1000\ mL}\right)\left(\frac{1\ mol\ H_2O}{1\ mol\ NaOH}\right)$$ ___0.0125 mol H_2O___

3. Using the average temperature change for reactions #1, 2, 3, how many joules of heat were released during the neutralization reactions?
   Show calculation set-ups:

   $$q = m(sp.ht.)\Delta T$$
   $$= (40.0\ mL)\left(\frac{1.04\ g}{1\ mL}\right)\left(\frac{4.184\ J}{g°C}\right)(4.2°C) = 730$$ ___730 J___

4. Using the average joules of heat calculated for these reactions and the moles H_2O produced for each trial, calculate the kJ/mole of H_2O formed.

   $$\frac{KJ}{mol\ H_2O} = \left(\frac{730\ J}{0.0125\ mol}\right)\left(\frac{1\ KJ}{1000\ J}\right) = 58.4\ KJ$$ ___58 $\frac{KJ}{mol}$___

5. Calculate the percent error for the experimental heat of neutralization?

   $$\left(\frac{58\frac{KJ}{mol} - 55.9\frac{KJ}{mol}}{55.9\frac{KJ}{mol}}\right)(100) =$$ ___+3.8%___

6. (a) How does the average temperature change in neutralization #4 compare to the average change determined for #1, 2, and 3?
   it is similar

   (b) How do you explain this when there was twice as much HCl (aq) involved in #4 vs. #1, 2, and 3?

   The added acid was in excess because the amt. of base was still the limiting reactant so there was no additional heat effect.

7. (a) How does the average temperature change for neutralization #5 compare with all the others?

   It went from 4.2 $\longrightarrow$ 7.8 so it was much greater (almost double)

   (b) Explain your answer.

   The additional base (double) neutralizes the excess acid which did not react before. This liberates more heat.

8. What is the kJ/mol H_2O formed for neutralization #5? Support your answer with calculation setups.

   $54.4\,\dfrac{KJ}{mol}$

   $$\text{mol } H_2O = (20.0\,\text{mL NaOH})\left(\frac{1.25\,\text{mol}}{1000\,\text{mL}}\right)\left(\frac{1\,\text{mol } H_2O}{1\,\text{mol NaOH}}\right) = 0.0250\,\text{mol } H_2O$$

   $$q = m(\text{sp.ht.})(\Delta T)$$

   $$(40.0\,\text{mL})\left(\frac{1.04\,\text{g}}{1\,\text{mL}}\right)\left(\frac{4.184\,J}{\text{g}^{\circ}C}\right)(7.8^{\circ}C) = 1360\,J$$

   $$\frac{KJ}{\text{mol } H_2O} = \left(\frac{1360\,J}{0.0250\,\text{mol } H_2O}\right)\left(\frac{1\,KJ}{1000\,J}\right) = 54.4\,\frac{KJ}{\text{mol } H_2O}$$

REPORT FOR EXPERIMENT 26

Distillation of Volatile Liquids

> ✓ Student results will be variable and will not exactly match the typical student data shown on the key as a guideline for grading.

Data Table

time, minutes	temp, °C ethanol	temp, °C H_2O		time, minutes	temp, °C 50% ethanol	temp, °C red wine	
0.0	34.0	34.0		0.0	34.0	34.0	
0.5	35.0	34.2		1.0	37.0	34.5	
1.0	37.0	34.5		2.0	39.0	35.0	
1.5	38.0	34.8		3.0	41.0	38.0	
2.0	44.0	35.5		4.0	43.0	42.0	
2.5	72.0	36.8		5.0	55.0	43.0	
3.0	76.0	38.0		6.0	75.0	45.0	
4.0	77.1	40.0		8.0	82.0	50.0	
5.0	77.5	43.0		10.0	83.0	58.0	
6.0	77.8	45.0		12.0	83.0	80.0	
7.0	78.0	49.0		14.0	84.0	87.0	
8.0	78.5	55.0		16.0	84.1	89.0	
9.0	78.0	60.0		18.0	84.5	91.5	
10.0	78.0	67.0		20.0	85.8	91.5	
11.0	77.8	85.0		22.0	85.8	93.0	
12.0	78.0	100.0		24.0	87.0	95.0	
13.0	78.1	100.1		26.0	89.0	97.0	
14.0	78.1	100.1		28.0	89.5	97.0	
15.0	78.0	100.1		30.0	92.5	97.0	
				32.0	93.8	97.0	
				34.0	96.0	97.0	
				36.0	97.2	97.0	
				38.0	100.0	97.0	
				40.0	101.0	97.0	
				42.0	101.0	97.0	
				44.0	101.0	97.0	

QUESTIONS AND PROBLEMS

1. The boiling points of three liquids are provided below.

   Acetone, bp = 56.2°C   Methanol, bp = 65°C   ⟨ **Ethylene glycol, bp = 198°C** ⟩

   Which liquid is the least volatile? (circle your choice)

   Which liquid has the weakest attractive forces between its molecules? (underline your choice)

2. If the drop of liquid on the thermometer disappears while the vapor is visibly condensing in the side arm, which process is proceeding faster in the flask, ⟨evaporation⟩ or condensation? (Circle your choice) How would you get the drop of liquid back?

   Turn down the heat (so evaporation will slow down).

3. Define boiling point.

   Temperature at which the vapor pressure of a liquid equals atmospheric pressure.

4. Why is the thermometer bulb placed above the liquid beside the side arm of the distilling flask and not in the liquid?

   To measure the temperature of the vapor that is going into the condenser. Measuring the temperature of the liquid does not give the temperature of the material distilling over.

5. What is the purpose of

   (a) the inner tube of the condenser?

   To provide a surface for removal of heat from the hot vapor and allow it to condense to a liquid.

   (b) the outer jacket of the condenser?

   To circulate water around the inner tube so hot vapor will condense into a liquid.

6. When the red wine is distilled, why does the distillate remain colorless throughout the full temperature range?

   The pigments in the solution (wine) are not very volatile so they remain in the distilling flask. The solution components which do distill are colorless.

7. Explain *briefly* how distillation can be used to purify seawater to make salt-free water.

   Seawater can be heated to distill the water as vapor which is condensed and collected as a liquid leaving behind the nonvolatile salt.

8. One team collected distillate from the wine in four 10 mL fractions in a graduated cylinder. What is the difference between the composition of the first 10 mL collected and the last 10 mL?

   The first 10 mL of distillate was mostly ethanol; the last 10 mL was mostly water.

   Explain this difference.

   The lower boiling ethanol distills first followed by the higher boiling water.

Use your graphs to answer the following questions:

9. (a) What is the experimental boiling point of ethanol? **about 78°C** water? **about 100°C**

   (b) What is the boiling point range for the 50% ethanol solution? **75°C–101°C**

10. Why did it take so much longer to reach the boiling point of water in the solutions of ethanol and water than in the pure water?

    In distilling a mixture of ethanol and water it takes time to remove the lower boiling ethanol before the boiling temperature of water can be reached.

Graph 1

Distillation of Ethanol vs. Water

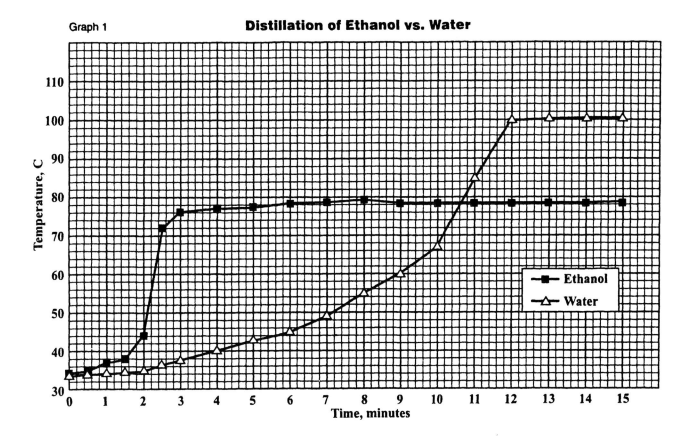

✓ These are typical computer graphs produced by a student from the data shown using the instructions in Study Aid 3.

Graph 2

Distillation of Ethanol/Water Solutions

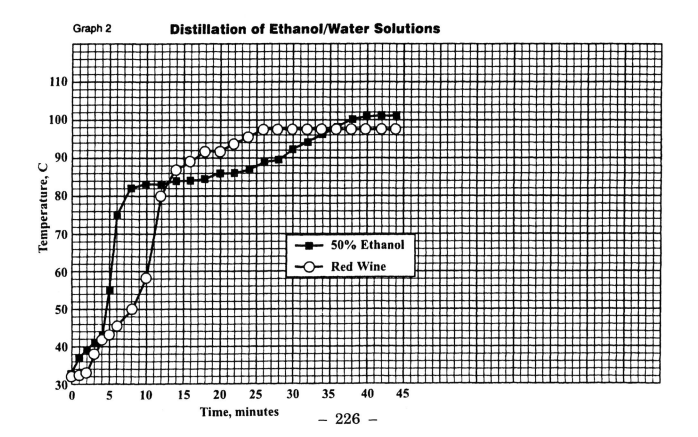

NAME KEY

SECTION ______ DATE __________

REPORT FOR EXPERIMENT 27

INSTRUCTOR __________

Hydrocarbons

A. Combustion

1. Describe the combustion characteristics of heptane and pentene.

Both burn rapidly with a yellow flame; pentene produces more black smoke (carbon) than heptane.

2. (a) Write a balanced equation to represent the complete combustion of heptane.

$$C_7H_{16} + 11\ O_2 \longrightarrow 7\ CO_2 + 8\ H_2O$$

(b) How many moles of oxygen are needed for the combustion of 1 mole of heptane in this reaction?

11 moles

B. and C. Reaction with Bromine and Potassium Permanganate

Data Table: Place an X in the column where a reaction was observed.

	Heptane (Saturated Hydrocarbon)	Pentene (Unsaturated Hydrocarbon)	Toluene (Aromatic Hydrocarbon)
Immediate reaction with Br_2 (*without* exposure to *light*)		X	
Slow reaction with Br_2 (*or only after exposure to light*)	X		
Reaction with $KMnO_4$		X	

1. Which of the three hydrocarbons reacted with bromine (without exposure to light)?

Pentene

2. Write an equation to illustrate how heptane reacts with bromine when the reaction mixture is exposed to sunlight.

$$C_7H_{16} + Br_2 \xrightarrow{\text{light}} C_7H_{15}Br + HBr$$

3. Write an equation to illustrate how pentene reacts with bromine. Assume the pentene is $CH_3CH_2CH=CHCH_3$ and use structural formulas.

$$CH_3CH_2CH=CHCH_3 + Br_2 \longrightarrow CH_3CH_2-\underset{\underset{Br}{|}}{CH}-\underset{\underset{Br}{|}}{CH}-CH_3$$

4. Which of the three hydrocarbons tested gave a positive Baeyer test?

_____**pentene**_____

D. Kerosene

1. (a) Did you observe any evidence of reaction with bromine before exposure to light? If so, describe.

   No. (There may be slight color fading with some kerosene samples.)

   (b) Did you observe any evidence of reaction with bromine after exposure to light? If so, describe.

   Yes. The color of the bromine faded (or disappeared).

2. Did you observe any evidence of reaction with potassium permanganate? If so, describe.

   No evidence of reaction.

3. Based on these tests (bromine and potassium permanganate), to which class of hydrocarbon does kerosene belong?

   Alkanes <u>or</u> saturated hydrocarbons

E. Acetylene

1. Describe the combustion characteristics of acetylene:

   (a) Tube 1.

   Yellow flame appears as tube is tilted back and forth. Considerable carbon (or soot) is formed during combustion.

   (b) Tube 2.

   Burns with very sooty (carbon) flame. (The entire test tube becomes coated with carbon.)

   (c) Tube 3.

   Tends to burn explosively (leaves very little carbon residue in the tube.)

2. Write an equation for the complete combustion of acetylene.

$$2 \, C_2H_2 + 5 \, O_2 \longrightarrow 4 \, CO_2 + 2 \, H_2O$$

F. Solubility Tests

1. Which of the three hydrocarbons tested are soluble in water?

None

2. From your observations what do you conclude about the density of hydrocarbons with respect to water?

They are less dense than water.

QUESTIONS AND PROBLEMS

1. Do you expect that acetylene would react with bromine without exposure to light? Explain your answer.

Yes. Pentene, an unsaturated hydrocarbon, reacted readily with bromine without exposure to light. Acetylene is unsaturated and should react like pentene.

2. Write condensed structural formulas for the three different isomers of pentane, all having the molecular formula C_5H_{12}. See Study Aid 6 for help if necessary.

$$CH_3-CH_2-CH_2-CH_2-CH_3 \qquad CH_3-\underset{\underset{\displaystyle CH_3}{|}}{CH}-CH_2-CH_3 \qquad CH_3-\underset{\underset{\displaystyle CH_3}{|}}{\overset{\overset{\displaystyle CH_3}{|}}{C}}-CH_3$$

3. Write condensed structural formulas for (a) ethene, (b) propene, and (c) the three different isomeric butenes (C_4H_8). See Study Aid 6 for help if necessary.

(a) $CH_2{=}CH_2$ (b) $CH_3CH{=}CH_2$

(c) $CH_3CH_2CH{=}CH_2$ $CH_3CH{=}CHCH_3$ $CH_2{=}\underset{\underset{\displaystyle CH_3}{|}}{C}-CH_3$

REPORT FOR EXPERIMENT 28

Alcohols, Esters, Aldehydes, and Ketones

A. Combustion of Alcohols

1. Compare the combustion characteristics of methyl, ethyl, and isopropyl alcohols, in terms of color and luminosity of their flames.

   Methyl alcohol—almost invisible blue flame or nonluminous flame.
   Ethyl alcohol—a more yellow flame than methyl alcohol (essentially nonluminous).
   Isopropyl alcohol—a still more yellow flame (than ethyl alcohol).

2. What type of flame would you predict for the combustion of amyl alcohol ($C_5H_{11}OH$)?
   Flame should be more yellow than in isopropyl alcohol in A.1. (indicative of incomplete combustion).

3. Write and balance the equation for the complete combustion of ethyl alcohol.

   $$C_2H_5OH + 3\,O_2 \longrightarrow 2\,CO_2 + 3\,H_2O$$

B. Oxidation of Alcohols

1. Oxidation with Potassium Permanganate

(a) Time required for oxidation of methyl alcohol by potassium permanganate:

   Tube 1: Alkaline solution. _____**Variable**_____

   Tube 2: Acid solution. _____**Variable**_____

   Tube 3: Neutral solution. _____**Variable**_____

(b) Time required for oxidation of isopropyl alcohol by potassium permanganate:

   Tube 1: Alkaline solution. _____**Variable**_____

   Tube 2: Acid solution. _____**Variable**_____

   Tube 3: Neutral solution. _____**Variable**_____

(c) Balance the equation for the oxidation of methyl alcohol:

   $$3\,CH_3OH + 2\,KMnO_4 \longrightarrow 3\,H_2C{=}O + 2\,KOH + 2\,H_2O + 2\,MnO_2$$

2. Oxidation with Copper(II) Oxide

(a) Write the equation for the oxidation reaction that occurred on the copper spiral when it was heated.

$$2\,Cu + O_2 \longrightarrow 2\,CuO$$

(b) What evidence of oxidation or reduction did you observe when the heated Cu spiral was lowered into methyl alcohol vapors?

The blackened surface of the spiral changed to a metallic copper appearance.

(c) Write and balance the oxidation-reduction equation between methyl alcohol and copper(II) oxide.

$$CH_3OH + CuO \longrightarrow Cu + H_2C{=}O + H_2O$$

C.   Formation of Esters

1. Describe the odor of:

(a) Ethyl acetate.

Fruity odor

(b) Isoamyl acetate.

Ripe banana odor

(c) Methyl salicylate.

Wintergreen odor

2. Write an equation to illustrate the formation of ethyl acetate from ethyl alcohol and acetic acid.

$$\underset{\displaystyle CH_3\overset{\textstyle O}{\overset{\|}{C}}-OH}{} + CH_3CH_2OH \xrightarrow[\Delta]{H_2SO_4} CH_3\overset{\textstyle O}{\overset{\|}{C}}-OCH_2CH_3 + H_2O$$

3. (a) The formula for isoamyl alcohol is $CH_3CH(CH_3)CH_2CH_2OH$. Write the formula for isoamyl acetate.

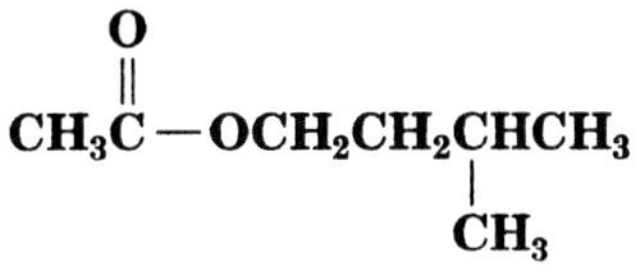

$$CH_3\overset{\overset{\displaystyle O}{\|}}{C}-OCH_2CH_2\underset{\underset{\displaystyle CH_3}{|}}{C}HCH_3$$

(b) The formula for salicylic acid is

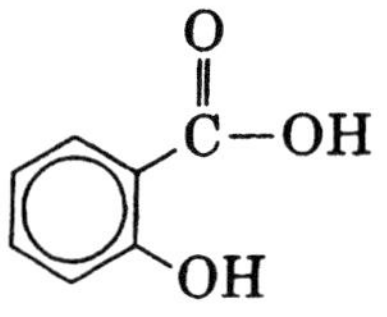

Write the formula for methyl salicylate.

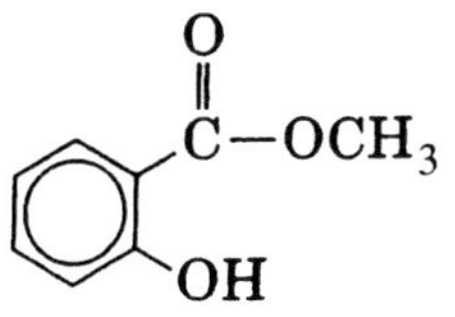

D.   Tollens Test for Aldehydes

1. How is a positive Tollens test recognized?

   By appearance of a silvery mirror in the test tube.

2. Which of the substances tested gave a positive Tollens test?

   Formaldehyde and glucose

3. Circle the formula(s) of the compounds listed that will give a positive Tollens test:

$$CH_3OH \qquad C_2H_5OH \qquad \boxed{CH_3\overset{\overset{\displaystyle H}{|}}{C}=O} \qquad CH_3\underset{\underset{\displaystyle O}{\|}}{C}CH_3 \qquad Na_2CO_3$$

4. Write the formula for the oxidation product formed from formaldehyde in the Tollens test.

   HCOONa or HCOOH

QUESTIONS AND PROBLEMS

1. There are four butyl alcohols of formula C_4H_9OH. Write their condensal structural formulas.

$CH_3-CH_2-CH_2-CH_2OH$ $CH_3-CH-CH_2OH$
                                                          |
                                                       CH_3

$CH_3-CH-CH_2-CH_3$ CH_3
              | |
            OH CH_3-C-OH
                                                          |
                                                       CH_3

2. Write the name of the ester that can be derived from the following pairs of acids and alcohols:

Alcohol	Acid	Ester
Methyl alcohol	Acetic acid	**Methyl acetate**
Ethyl alcohol	Formic acid	**Ethyl formate**
Isopropyl alcohol	Butyric acid	**Isopropyl butyrate**

3. Write condensed structural formular for all the five carbon aldehydes and hetones. The molecular formula is $C_5H_{10}O$.

Aldehydes

$$\overset{\displaystyle O}{\underset{\displaystyle ||}{}}$$
$CH_3CH_2CH_2CH_2C-H$

$$\overset{O}{||}$$
CH_3CH_2CHC-H
              |
            CH_3

$$\overset{O}{||}$$
CH_3CHCH_2C-H
       |
     CH_3

CH_3 O
   | ||
CH_3C-C-H
   |
CH_3

Ketones

$CH_3CH_2CH_2CCH_3$
                        ||
                        O

$CH_3CH_2CCH_2CH_3$
                ||
                O

      CH_3
         |
CH_3CHCCH_3
                ||
                O

EXERCISE 1

Significant Figures and Exponential Notation

1. How many significant figures are in each of the following numbers?

   (a) 7.42 ____**3**____ (b) 4.6 ____**2**____ (c) 3.40 ____**3**____ (d) 26,000 ___**2**___

   (e) 0.088 ___**2**___ (f) 0.0034 __**2**__ (g) 0.0230 ___**3**___ (h) 0.3080 ___**4**___

2. Write each of the following numbers in proper exponential notation:

   (a) 423   (a) ____**4.23 × 10²**____

   (b) 0.032   (b) ____**3.2 × 10⁻²**____

   (c) 8,300   (c) ____**8.3 × 10³**____

   (d) 302.0   (d) ____**3.020 × 10²**____

   (e) 12,400,000   (e) ____**1.24 × 10⁷**____

   (f) 0.0007   (f) _____**7 × 10⁻⁴**_____

3. How many significant figures should be in the answer to each of the following calculations?

   (a) 17.10   (b) 57.826   (a) _____**4**_____
       + 0.77   − 9.4
       _____   _____   (b) _____**3**_____

                                               (c) _____**3**_____

   (c) $12.4 \times 2.82 =$ (d) $6.4 \times 3.1416 =$ (d) _____**2**_____

                                               (e) _____**4**_____

   (e) $\dfrac{0.5172}{0.2742} =$ (f) $\dfrac{0.0172}{4.36} =$ (f) _____**3**_____

   (g) $\dfrac{5.82 \times 760. \times 425}{723 \times 273} =$ (h) $\dfrac{0.92 \times 454 \times 5.620}{22.4} =$ (g) _____**3**_____

                                               (h) _____**2**_____

4. For each of these problems, complete the answer with a 10 raised to the proper power. Note that each answer is expressed to the correct number of significant figures.

(a) $2.71 \times 10^4 \times 2.0 \times 10^2 = 5.4 \times$ _____   (a) _____ **10^6** _____

(b) $\dfrac{4.523 \times 10^4}{2.71 \times 10^2} = 1.67 \times$ _____   (b) _____ **10^2** _____

(c) $4.8 \times 10^4 \times 3.5 \times 10^4 = 1.7 \times$ _____   (c) _____ **10^9** _____

(d) $\dfrac{1.64 \times 10^4}{1.2 \times 10^2} = 1.4 \times$ _____   (d) _____ **10^{-6}** _____

(e) $\dfrac{4.70 \times 10^2}{8.42 \times 10^5} = 5.58 \times$ _____   (e) _____ **10^{-4}** _____

5. Solve each of the following problems, expressing each answer to the proper number of significant figures. Use exponential notation for (c), (d), and (e).

(a)   1.842   (b) 714.3

      45.21   − 28.52   (a) _____ **84.60** _____

    + 37.55   (b) _____ **685.8** _____

(c) $2.83 \times 10^3 \times 7.55 \times 10^7 =$ (c) _____ **2.14×10^{11}** _____

(d) $4.4 \times 5{,}280 =$ (d) _____ **2.3×10^4** _____

(e) $\dfrac{7.07 \times 10^{-4} \times 6.51 \times 10^{-2}}{2.92 \times 10^4} =$ (e) _____ **1.58×10^{-9}** _____

Answers

1. (a) 3, (b) 2, (c) 3, (d) 2, (e) 2, (f) 2, (g) 3, (h) 4.

2. (a) 4.23×10^2, (b) 3.2×10^{-2}, (c) 8.3×10^3, (d) 3.020×10^2, (e) 1.24×10^7, (f) 7×10^{-4}.

3. (a) 4, (b) 3, (c) 3, (d) 2, (e) 4, (f) 3, (g) 3, (h) 2.

4. (a) 10^6, (b) 10^2, (c) 10^9, (d) 10^{-6}, (e) 10^{-4}.

5. (a) 84.60, (b) 685.8, (c) 2.14×10^{11}, (d) 2.3×10^4, (e) 1.58×10^{-9}.

EXERCISE 2

Measurements

For each of the following problems, show your calculation setup. In both your setup and answer, show units and follow the rules of significant figures. See Experiment 2 and the appendixes for any needed formulas or conversion factors.

1. Convert 78°F to degrees Celsius.

$$°C = \frac{(78°F - 32)}{1.8} = \frac{46}{1.8} = 26°C$$

26°C

2. Convert −13°C to degrees Fahrenheit.

$$°F = 1.8(-13) + 32 = -23 + 32 = +9°F$$

+9°F

3. An object weighs 8.22 lbs. What is the mass in grams?

$$(8.22 \text{ lb})\left(\frac{453.6 \text{ g}}{\text{lb}}\right) = 3.73 \times 10^3 \text{ g}$$

3.73×10^3 g

4. A stick is 12.0 cm long. What is the length in inches?

$$(12.0 \text{ cm})\left(\frac{1 \text{ in.}}{2.54 \text{ cm}}\right) = 4.72 \text{ in.}$$

4.72 in.

5. The water in a flask measures 423 mL. How many quarts is this?

$$(423 \text{ mL})\left(\frac{1 \text{ qt}}{946 \text{ mL}}\right) = 0.447 \text{ qt}$$

0.477 qt

6. A piece of lumber measures 98.4 cm long. What is its length in:

(a) Millimeters? $(98.4 \text{ cm})\left(\dfrac{10 \text{ mm}}{\text{cm}}\right) = 984 \text{ mm}$

984 mm

(b) Feet? $(98.4 \text{ cm})\left(\dfrac{1 \text{ in.}}{2.54 \text{ cm}}\right)\left(\dfrac{1 \text{ ft}}{12 \text{ in.}}\right) = 3.23 \text{ ft}$

3.23 ft

7. A block is found to have a volume of 35.3 cm³. Its mass is 31.7 g. Calculate the density of the block.

$$d = \frac{m}{v} = \frac{31.7\,g}{35.3\,cm^3} = 0.898\,g/cm^3$$

_____
0.898 g/cm³

8. A graduated cylinder was filled to 25.0 mL with liquid. A solid object weighing 73.5 g was immersed in the liquid, raising the liquid level to 43.9 mL. Calculate the density of the solid object.

43.9 mL   filled
25.0 mL   empty $d = \dfrac{m}{v}$ $\dfrac{73.5\,g}{18.9\,mL} = 3.89\,g/mL$
18.9 mL   liquid

_____
3.89 g/mL

9. The density of the liquid in Problem 8 is 0.874 g/mL. What is the mass of the liquid in the graduated cylinder.

$$(25.0\,mL)\left(\frac{0.874\,g}{mL}\right) = 21.9\,g$$

_____
21.9 g

10. How many joules of heat are absorbed by 500.0 g of water when its temperature increases from 20.0°C to 80.0°C? (sp. ht. water = 1.00 cal/g°C)

$$(500.0\,g)\left(\frac{1.00\,cal}{g°C}\right)\left(\frac{4.184\,J}{1\,cal}\right)(80.0°C - 20.0°C) =$$

_____
1.26 × 10⁵ J

11. A beaker contains 421 mL of water. The density of the water is 1.00 g/mL. Calculate:

(a) The volume of the water in liters.

$$(421\,mL)\left(\frac{1\,L}{1000\,mL}\right) = 0.421\,L$$

_____
0.421 L

(b) The mass of the water in grams.

$$(421\,mL)\left(\frac{1.00\,g}{mL}\right) = 421\,L$$

_____
421 g

12. The density of carbon tetrachloride, CCl₄, is 1.59 g/mL. Calculate the volume of 100.0 g of CCl₄.

$$(100.0\,g)\left(\frac{1\,mL}{1.59\,g}\right) = 62.9\,mL$$

_____
62.9 mL

EXERCISE 3

Names and Formulas 1

Give the names of the following compounds:

1. NaCl — **Sodium chloride**

2. $AgNO_3$ — **Silver nitrate**

3. $BaCrO_4$ — **Barium chromate**

4. $Ca(OH)_2$ — **Calcium hydroxide**

5. $ZnCO_3$ — **Zinc carbonate**

6. Na_2SO_4 — **Sodium sulfate**

7. Al_2O_3 — **Aluminum oxide**

8. $CdBr_2$ — **Cadmium bromide**

9. KNO_2 — **Potassium nitrite**

10. $Fe(NO_3)_3$ — (a) **Iron(III) nitrate** — (b) **Ferric nitrate**

11. $(NH_4)_3PO_4$ — **Ammonium phosphate**

12. $KClO_3$ — **Potassium chlorate**

13. MgS — **Magnesium sulfide**

14. $Cu_2C_2O_4$ — **Copper(I) oxalate**

Give the formulas of the following compounds:

1. Barium chloride   1. _____ $BaCl_2$ _____

2. Zinc fluoride   2. _____ ZnF_2 _____

3. Lead(II) iodide   3. _____ PbI_2 _____

4. Ammonium hydroxide   4. _____ NH_4OH _____

5. Potassium chromate   5. _____ K_2CrO_4 _____

6. Bismuth(III) chloride   6. _____ $BiCl_3$ _____

7. Magnesium perchlorate   7. _____ $Mg(ClO_4)_2$ _____

8. Copper(II) sulfate   8. _____ $CuSO_4$ _____

9. Iron(III) chloride   9. _____ $FeCl_3$ _____

10. Calcium cyanide   10. _____ $Ca(CN)_2$ _____

11. Copper(I) sulfide   11. _____ Cu_2S _____

12. Silver carbonate   12. _____ Ag_2CO_3 _____

13. Cadmium hypochlorite   13. _____ $Cd(ClO)_2$ _____

14. Sodium bicarbonate   14. _____ $NaHCO_3$ _____

15. Aluminum acetate   15. _____ $Al(C_2H_3O_2)_3$ _____

16. Nickel(II) phosphate   16. _____ $Ni_3(PO_4)_2$ _____

17. Sodium sulfite   17. _____ Na_2SO_3 _____

18. Tin(IV) oxide   18. _____ SnO_2 _____

EXERCISE 4

Names and Formulas II

Give the names of the following compounds:

1. $(NH_4)_2S$ _____**Ammonium sulfide**_____

2. NiF_2 _____**Nickel(II) fluoride**_____

3. $Sb(ClO_3)_3$ _____**Antimony(III) chlorate**_____

4. $HgCl_2$ _____**Mercury(II) chloride**_____

5. $H_2SO_4(aq)$ _____**Sulfuric acid**_____

6. $CrBr_3$ _____**Chromium(III) bromide**_____

7. Cu_2CO_3 (a) _____**Copper(I) carbonate**_____

                    (b) _____**Cuprous carbonate**_____

8. $K_2Cr_2O_7$ _____**Potassium dichromate**_____

9. $FeSO_4$ (a) _____**Iron(II) sulfate**_____

                    (b) _____**Ferrous sulfate**_____

10. $AgC_2H_3O_2$ _____**Silver acetate**_____

11. HCl _____**Hydrogen chloride**_____

12. $HCl(aq)$ _____**Hydrochloric acid**_____

13. $KBrO_3$ _____**Potassium bromate**_____

14. $Cd(ClO_2)_2$ _____**Cadmium chlorite**_____

EXERCISE 4 (continued) **NAME** KEY

Give the formulas of the following compounds:

1. Sodium oxalate	1. $Na_2C_2O_4$
2. Manganese(II) iodate	2. $Mn(IO_3)_2$
3. Zinc nitrite	3. $Zn(NO_2)_2$
4. Potassium permanganate	4. $KMnO_4$
5. Titanium(IV) bromide	5. $TiBr_4$
6. Sodium arsenate	6. Na_3AsO_4
7. Manganese(IV) sulfide	7. MnS_2
8. Bismuth(III) arsenate	8. $BiAsO_4$
9. Sodium peroxide	9. Na_2O_2
10. Magnesium bicarbonate	10. $Mg(HCO_3)_2$
11. Lead(II) acetate	11. $Pb(C_2H_3O_2)_2$
12. Phosphoric acid	12. $H_3PO_4(aq)$
13. Nitric acid	13. $HNO_3(aq)$
14. Acetic acid	14. $HC_2H_3O_2(aq)$
15. Arsenic(III) iodide	15. AsI_3
16. Ammonium thiocyanate	16. NH_4SCN
17. Cobalt(II) chlorite	17. $Co(ClO_2)_2$
18. Stannous fluoride	18. SnF_2

EXERCISE 5

Names and Formulas III

Give the names of the following compounds:

1. CO_2 — **Carbon dioxide**

2. H_2O_2 — **Hydrogen peroxide**

3. $Ni(MnO_4)_2$ — **Nickel(II) permanganate**

4. $Co_3(AsO_4)_2$ — **Cobalt(II) arsenate**

5. KCN — **Potassium cyanide**

6. Sb_2O_5 — **Antimony(V) oxide**

7. BaH_2 — **Barium hydride**

8. $NaHSO_3$ — **Sodium bisulfite or Sodium hydrogen sulfite**

9. $As(NO_2)_5$ — **Arsenic(V) nitrite**

10. $KSCN$ — **Potassium thiocyanate**

11. Ag_2CO_3 — **Silver carbonate**

12. CrF_3 — **Chromium(III) fluoride**

13. SnS_2 (a) — **Tin(IV) sulfide**

    (b) — **Stannic sulfide**

14. $H_2SO_3(aq)$ — **Sulfurous acid**

15. HgC_2O_4 — **Mercury(II) oxalate or Mecuric oxalate**

16. $Pb(HCO_3)_2$ — **Lead(II) bicarbonate or Lead(II) hydrogen carbonate**

17. $Cu(OH)_2$ — **Copper(II) hydroxide**

Give the formulas of the following substances:

	Substance		Formula
1.	Ammonium hydrogen carbonate	1.	NH_4HCO_3
2.	Hydrogen sulfide	2.	H_2S
3.	Barium hydroxide	3.	$Ba(OH)_2$
4.	Carbon tetrachloride	4.	CCl_4
5.	Nickel(II) perchlorate	5.	$Ni(ClO_4)_2$
6.	Lead(II) nitrate	6.	$Pb(NO_3)_2$
7.	Sulfur dioxide	7.	SO_2
8.	Carbonic acid	8.	H_2CO_3
9.	Copper(I) carbonate	9.	Cu_2CO_3
10.	Calcium cyanide	10.	$Ca(CN)_2$
11.	Arsenic(III) oxide	11.	As_2O_3
12.	Silver dichromate	12.	$Ag_2Cr_2O_7$
13.	Nitrous acid	13.	$HNO_2(aq)$
14.	Copper(II) bromide	14.	$CuBr_2$
15.	Ammonia	15.	NH_3
16.	Chlorine	16.	Cl_2
17.	Chromium(III) sulfite	17.	$Cr_2(SO_3)_3$
18.	Chloric acid	18.	$HClO_3(aq)$
19.	Barium arsenate	19.	$Ba_3(AsO_4)_2$
20.	Manganese(IV) chloride	20.	$MnCl_4$
21.	Carbon disulfide	21.	CS_2
22.	Aluminum fluoride	22.	AlF_3

EXERCISE 6

Equation Writing and Balancing I

Balance the following equations:

1. $2\,Mg + O_2 \xrightarrow{\Delta} 2\,MgO$

2. $2\,KClO_3 \xrightarrow{\Delta} 2\,KCl + 3\,O_2$

3. $3\,Fe + 2\,O_2 \xrightarrow{\Delta} Fe_3O_4$

4. $Mg + 2\,HCl \longrightarrow MgCl_2 + H_2$

5. $2\,Na + 2\,H_2O \longrightarrow 2\,NaOH + H_2$

Beneath each word equation write the formula equation and balance it. Remember that oxygen and hydrogen are diatomic molecules.

1. Sulfur + Oxygen $\xrightarrow{\Delta}$ Sulfur dioxide

   $S + O_2 \longrightarrow SO_2$

2. Zinc + Sulfuric acid $\longrightarrow$ Zinc sulfate + Hydrogen

   $Zn + H_2SO_4 \longrightarrow ZnSO_4 + H_2$

3. Carbon + Oxygen $\xrightarrow{\Delta}$ Carbon dioxide

   $C + O_2 \longrightarrow CO_2$

4. Hydrogen + Oxygen $\xrightarrow{\Delta}$ Water

   $2\,H_2 + O_2 \longrightarrow 2\,H_2O$

5. Aluminum + Hydrochloric acid $\longrightarrow$ Aluminum chloride + Hydrogen

   $2\,Al + 6\,HCl \longrightarrow 2\,AlCl_3 + 3\,H_2$

Balance the following equations:

1. $N_2 + 3\ H_2 \xrightarrow{\Delta} 2\ NH_3$

2. $CoCl_2 \cdot 6\ H_2O \xrightarrow{\Delta} CoCl_2 + 6\ H_2O$

3. $3\ Fe + 4\ H_2O \xrightarrow{\Delta} Fe_3O_4 + 4\ H_2$

4. $2\ F_2 + 2\ H_2O \xrightarrow{\Delta} 4\ HF + O_2$

5. $2\ Pb(NO_3)_2 \xrightarrow{\Delta} 2\ PbO + 4\ NO + 3\ O_2$

Beneath each word equation write and balance the formula equation. Oxygen, hydrogen, and bromine are diatomic molecules

1. Aluminum + Oxygen $\xrightarrow{\Delta}$ Aluminum oxide

   $4\ Al + 3\ O_2 \longrightarrow 2\ Al_2O_3$

2. Potassium + Water $\longrightarrow$ Potassium hydroxide + Hydrogen

   $2\ K + 2\ H_2O \longrightarrow 2\ KOH + H_2$

3. Arsenic(III) oxide + Hydrochloric acid $\longrightarrow$ Arsenic(III) chloride + Water

   $As_2O_3 + 6\ HCl \longrightarrow 2\ AsCl_3 + 3\ H_2O$

4. Phosphorus + Bromine $\longrightarrow$ Phosphorus tribromide

   $2\ P + 3\ Br_2 \longrightarrow 2\ PBr_3$

5. Sodium bicarbonate + Nitric acid $\longrightarrow$ Sodium nitrate + Water + Carbon dioxide

   $NaHCO_3 + HNO_3 \longrightarrow NaNO_3 + H_2O + CO_2$

EXERCISE 7

Equation Writing and Balancing II

Complete and balance the following double displacement reaction equations (assume all reactions will go):

1. $NaCl +$ $AgNO_3 \longrightarrow$ **AgCl(s) + NaNO$_3$**

2. $BaCl_2 +$ $H_2SO_4 \longrightarrow$ **BaSO$_4$(s) + 2 HCl**

3. $NaOH +$ $HCl \longrightarrow$ **NaCl + H$_2$O**

4. $Na_2CO_3 +$ 2 $HCl \longrightarrow$ **2 NaCl + H$_2$O + CO$_2$(g)**

5. $H_2SO_4 +$ 5 $NH_4OH \longrightarrow$ **(NH$_4$)$_2$SO$_4$ + 2 H$_2$O**

6. $FeCl_3 +$ 3 $NH_4OH \longrightarrow$ **Fe(OH)$_3$(s) + 3 NH$_4$Cl**

7. $Na_2SO_3 +$ 2 $HCl \longrightarrow$ **2 NaCl + H$_2$O + SO$_2$(g)**

8. $K_2CrO_4 +$ $Pb(NO_3)_2 \longrightarrow$ **2 KNO$_3$ + PbCrO$_4$(s)**

9. $NaC_2H_3O_2 +$ $HCl \longrightarrow$ **NaCl + HC$_2$H$_3$O$_2$**

10. $NaOH +$ $NH_4NO_3 \longrightarrow$ **NaNO$_3$ + H$_2$O + NH$_3$(g)**

11.   2 $BiCl_3 +$ 3 $H_2S \longrightarrow$ **Bi$_2$S$_3$(s) + 6 HCl**

12. $K_2C_2O_4 +$ 2 $HCl \longrightarrow$ **2 KCl + H$_2$C$_2$O$_4$**

13.   2 $H_3PO_4 +$ 3 $Ca(OH)_2 \longrightarrow$ **Ca$_3$(PO$_4$)$_2$(s) + 6 H$_2$O**

14. $(NH_4)_2CO_3 +$ 2 $HNO_3 \longrightarrow$ **2 NH$_4$NO$_3$ + H$_2$O + CO$_2$(g)**

15. $K_2CO_3 +$ $NiBr_2 \longrightarrow$ **NiCO$_3$(s) + 2 KBr**

Complete and balance the following equations. (Combination, 1–4; Decomposition, 5–8; Single displacement, 9–12; Double displacement, 13–16.)

1. $2\,K + Cl_2 \longrightarrow 2\,KCl$

2. $2\,Zn + O_2 \longrightarrow 2\,ZnO$

3. $BaO + H_2O \longrightarrow Ba(OH)_2$

4. $SO_3 + H_2O \longrightarrow H_2SO_4$

5. $MgCO_3 \xrightarrow{\Delta} MgO + CO_2(g)$

6. $NH_4OH \xrightarrow{\Delta} NH_3(g) + H_2O$

7. $Mn(ClO_3)_2 \xrightarrow{\Delta} MnCl_2 + 3\,O_2(g)$

8. $2\,HgO \xrightarrow{\Delta} 2\,Hg + O_2(g)$

9. $Ni + 2\ HCl \longrightarrow NiCl_2 + H_2(g)$

10. $Pb + 2\ AgNO_3 \longrightarrow Pb(NO_3)_2 + 2\,Ag(s)$

11. $Cl_2 + 2\ NaI \longrightarrow I_2 + 2\,NaCl$

12. $2\,Al + 3\ CuSO_4 \longrightarrow Al_2(SO_4)_3 + 3\,Cu(s)$

13. $3\,KOH + H_3PO_4 \longrightarrow K_3PO_4 + 3\,H_2O$

14. $Na_2C_2O_4 + CaCl_2 \longrightarrow 2\,NaCl + CaC_2O_4(s)$

15. $(NH_4)_2SO_4 + 2\ KOH \longrightarrow K_2SO_4 + 2\,NH_3(g) + 2\,H_2O$

16. $ZnCl_2 + (NH_4)_2S \longrightarrow ZnS(s) + 2\,NH_4Cl$

EXERCISE 8

Equation Writing and Balancing III

For each of the following situations, write and balance the formula equation for the reaction that occurs.

1. A strip of zinc is dropped into a test tube of hydrochloric acid.
   $$Zn + 2\,HCl \longrightarrow ZnCl_2 + H_2(g)$$

2. Hydrogen peroxide decomposes in the presence of manganese dioxide.
   $$2\,H_2O_2 \xrightarrow{MnO_2} 2\,H_2O + O_2(g)$$

3. Copper(II) sulfate pentahydrate is heated to drive off the water of hydration.
   $$CuSO_4 \cdot 5\,H_2O \xrightarrow{\Delta} CuSO_4 + 5\,H_2O(g)$$

4. A piece of sodium is dropped into a beaker of water.
   $$2\,Na + 2\,H_2O \longrightarrow 2\,NaOH + H_2(g)$$

5. A piece of limestone (calcium carbonate) is heated in a Bunsen burner flame.
   $$CaCO_3 \xrightarrow{\Delta} CaO + CO_2(g)$$

6. A piece of zinc is dropped into a solution of silver nitrate.
   $$Zn + 2\,AgNO_3 \longrightarrow Zn(NO_3)_2 + 2\,Ag(s)$$

7. Hydrochloric acid is added to a sodium carbonate solution.
   $$2\,HCl + Na_2CO_3 \longrightarrow 2\,NaCl + H_2O + CO_2(g)$$

8. Potassium chlorate is heated in the presence of manganese dioxide.
   $$2\,KClO_3 \xrightarrow[MnO_2]{\Delta} 2\,KCl + 3\,O_2(g)$$

9. Hydrogen gas is burned in air.
   $$2\,H_2 + O_2 \longrightarrow 2\,H_2O$$

10. Sulfuric acid solution is reacted with sodium hydroxide solution.
    $$H_2SO_4 + 2\,NaOH \longrightarrow Na_2SO_4 + 2\,H_2O$$

EXERCISE 9

Graphical Representation of Data

A. From the figure at the right, read values for the following:

1. The vapor pressure of ethyl ether at 20°C.

    _____**about 450 torr**_____

2. The temperature at which ethyl chloride has a vapor pressure of 620 torr.

    _____**about 8°C**_____

3. The temperature at which ethyl alcohol has the pressure that ethyl chloride has at 2°C.

    _____**about 68°C**_____

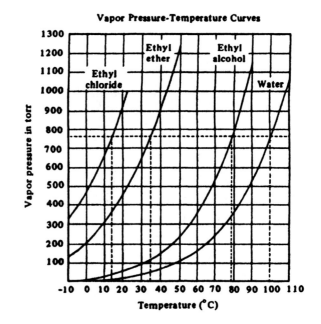

Vapor Pressure-Temperature Curves

B. Plotting Graphs

1. Plot the following pressure-temperature data for a gas on the graph below. Draw the best possible straight line through the data. Provide temperature and pressure scales.

Temperature, °C	0	20	40	60	80	100
Pressure, torr	586	628	655	720	757	800

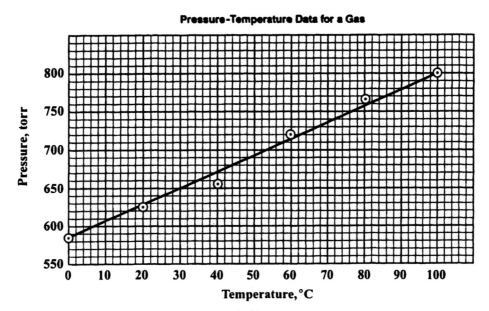

Pressure-Temperature Data for a Gas

2. (a) Study the data given below; (b) determine suitable scales for pressure and for volume and mark these scales on the graph; (c) plot eight points on the graph; (d) draw the best possible line through these points; (e) place a suitable title at the top of the graph.

Pressure-volume data for a gas

Volume, mL	10.70	7.64	5.57	4.56	3.52	2.97	2.43	2.01
Pressure, torr	250	350	480	600	760	900	1100	1330

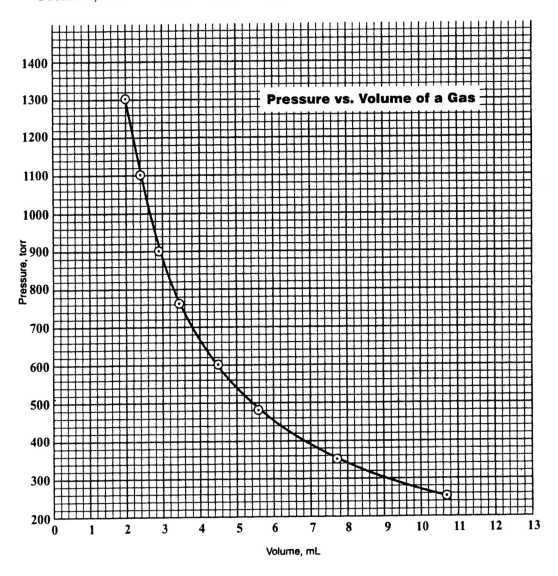

Read from your graph:

(a) The pressure at 10.0 mL ____**about 271 torr**____

(b) The volume at 700 torr ____**about 3.8 mL**____

EXERCISE 10

Moles

Show calculation setups and answers for all problems.

1. Find the molar mass of (a) nitric acid, HNO_3; (b) potassium bicarbonate, $KHCO_3$; and (c) Nickel(II) nitrate, $Ni(NO_3)_2$.

HNO_3	$KHCO_3$	$Ni(NO_3)_2$
1(1.008) = 1.008	1(39.10) = 39.10	1(58.69) = 58.69
1(14.01) = 14.01	1(1.008) = 1.008	2(14.01) = 28.02
3(16.00) = 48.00	1(12.01) = 12.01	6(16.00) = 96.00
63.02	3(16.00) = 48.00	182.7
	100.1	

(a) _____ 63.02 g/mol _____

(b) _____ 100.1 g/mol _____

(c) _____ 182.7 g/mol _____

2. A sample of mercury(II) bromide, $HgBr_2$, weighs 8.65 g. How many moles are in this sample?

$$(8.65\ HgBr_2)\left(\frac{1\ mol}{360.4\ g}\right) = 0.0240\ mol$$

_____ 0.0240 mol $HgBr_2$ _____

3. What is the mass of 0.45 mol of ammonium sulfate, $(NH_4)_2SO_4$?

$$(0.45\ mol\ (NH_4)_2SO_4)\left(\frac{132.2\ g}{1\ mol}\right) = 59\ g$$

_____ 59 g _____

4. How many molecules are contained in 6.53 mol of nitrogen gas, N_2?

$$(6.53\ mol\ N_2)\left(\frac{6.022 \times 10^{23}\ molecules}{1\ mol}\right) = 3.93 \times 10^{24}\ molecules$$

_____ 3.93×10^{24} molecules _____

5. Calculate the percent composition by mass of calcium sulfite, $CaSO_3$.

$CaSO_3 = 120.2\ g/mol$

$$\%\ Ca = \left(\frac{40.08\ g}{120.2\ g}\right)(100) =$$

$$\%\ S = \left(\frac{32.07\ g}{120.2\ g}\right)(100) =$$

$$\%\ O = \left(\frac{3(16.00\ g)}{120.2\ g}\right)(100) =$$

Ca _____ 33.34% _____

S _____ 26.68% _____

O _____ 39.93% _____

6. An organic compound is analyzed and found to be carbon 51.90%, hydrogen 9.80%, and chlorine 38.30%. What is the empirical formula of this compound?

C $(51.90\,\text{g})\left(\dfrac{1\,\text{mol}}{12.01\,\text{g}}\right) = 4.321\,\text{mol}$ $C_{\frac{4.321}{1.080}}$ $H_{\frac{9.72}{1.080}}$ $Cl_{\frac{1.080}{1.080}}$

H $(9.80\,\text{g})\left(\dfrac{1\,\text{mol}}{1.008\,\text{g}}\right) = 9.72\,\text{mol}$ C_4H_9Cl

Cl $(38.30\,\text{g})\left(\dfrac{1\,\text{mol}}{35.45\,\text{g}}\right) = 1.080\,\text{mol}$

7. A sample of oxygen gas, O_2, weighs 28.4 g. How many molecules of O_2 and how many atoms of O are present in this sample?

$(28.4\,\text{g}\,O_2)\left(\dfrac{1\,\text{mol}}{32.00\,\text{g}}\right)\left(\dfrac{6.022 \times 10^{23}\,\text{molecules}}{\text{mol}}\right) =$

$(5.34 \times 10^{23}\,\text{molecules}\,O_2)\left(\dfrac{2\,\text{atoms O}}{\text{molecule}\,O_2}\right) =$

_____ 5.34×10^{23} _____ molecules of O_2

_____ 1.07×10^{24} _____ atoms of O

8. A mixture of sand and salt is found to be 48 percent NaCl by mass. How many moles of NaCl are in 74 g of this mixture?

$(74\,\text{g mixture})\left(\dfrac{48\,\text{g NaCl}}{100.\,\text{g mixture}}\right)\left(\dfrac{1\,\text{mol}}{58.44\,\text{g}}\right) = 0.61\,\text{mol}$

_____ 0.61 mol NaCl _____

9. What is the mass of 2.6×10^{23} molecules of ammonia, NH_3?

$(2.6 \times 10^{23}\,\text{molecules}\,NH_3)\left(\dfrac{1\,\text{mol}}{6.022 \times 10^{23}\,\text{molecules}}\right)\left(\dfrac{17.03\,\text{g}}{\text{mol}}\right) =$

_____ 7.4 g NH_3 _____

10. A water solution of sulfuric acid has a density of 1.67 g/mL and is 75 percent H_2SO_4 by mass. How many moles of H_2SO_4 are contained in 400. mL of this solution?

$(400.\,\text{mL sol.})\left(\dfrac{1.67\,\text{g solution}}{1\,\text{mL sol.}}\right)\left(\dfrac{75\,\text{g}\,H_2SO_4}{100.\,\text{g sol.}}\right)\left(\dfrac{1\,\text{mol}}{98.09\,\text{g}\,H_2SO_4}\right) =$

_____ 5.1 mol H_2SO_4 _____

EXERCISE 11

Stoichiometry I

Show calculation setups and answers for all problems.

1. Use the equation given to solve the following problems:

$$Na_3PO_4 + 3\,AgNO_3 \longrightarrow Ag_3PO_4 + 3\,NaNO_3$$

(a) How many moles of Na_3PO_4 would be required to react with 1.0 mol of $AgNO_3$?

$$(1.0\,mol\,AgNO_3)\left(\frac{1\,mol\,Na_3PO_4}{3\,mol\,AgNO_3}\right) =$$

__0.33 mol Na_3PO_4__

(b) How many moles of $NaNO_3$ can be produced from 0.50 mol of Na_3PO_4?

$$(0.50\,mol\,Na_3PO_4)\left(\frac{3\,mol\,NaNO_3}{1\,mol\,Na_3PO_4}\right) =$$

__1.5 mol $NaNO_3$__

(c) How many grams of Ag_3PO_4 can be produced from 5.00 g of Na_3PO_4?

$$(5.00\,g\,Na_3PO_4)\left(\frac{1\,mol\,Na_3PO_4}{163.9\,g\,Na_3PO_4}\right)\left(\frac{1\,mol\,Ag_3PO_4}{3\,mol\,Na_3PO_4}\right)\left(\frac{418.7\,g\,Ag_3PO_4}{mol\,Ag_3PO_4}\right) =$$

__12.8 g Ag_3PO_4__

(d) If you have 9.44 g of Na_3PO_4, how many grams of $AgNO_3$ will be needed for complete reaction?

$$(9.44\,g\,Na_3PO_4)\left(\frac{1\,mol\,Na_3PO_4}{163.9\,g\,Na_3PO_4}\right)\left(\frac{3\,mol\,AgNO_3}{1\,mol\,Na_3PO_4}\right)\left(\frac{169.9\,g\,AgNO_3}{mol\,AgNO_3}\right) \times$$

__29.4 g $AgNO_3$__

(e) When 25.0 g of $AgNO_3$ are reacted with excess Na_3PO_4, 18.7 g of Ag_3PO_4 are produced. What is the percentage yield of Ag_3PO_4?

$$(25.0\,g\,Na_3PO_4)\left(\frac{1\,mol\,AgNO_3}{169.9\,g\,AgNO_3}\right)\left(\frac{1\,mol\,Ag_3PO_4}{3\,mol\,AgNO_3}\right)\left(\frac{418.7\,g\,Ag_3PO_4}{mol\,Ag_3PO_4}\right) = 20.5\,g\,Ag_3O_4$$
(theoretical yield)

$$\frac{18.7\,g}{20.5\,g} \times 100 = 91.2\%\ \text{yield}$$

__91.2%__

2. Use the equation given to solve the following problems:

$$2\,KMnO_4 + 16\,HCl \longrightarrow 5\,Cl_2 + 2\,KCl + 2\,MnCl_2 + 8\,H_2O$$

(a) How many moles of HCl are required to react with 35 g of $KMnO_4$?

$$(35\,g\,KMnO_4)\left(\frac{1\,mol\,KMnO_4}{158.0\,g\,KMnO_4}\right)\left(\frac{16\,mol\,HCL}{2\,mol\,KMnO_4}\right) =$$

<u>   1.8 mol HCl   </u>

(b) How many Cl_2 molecules will be produced using 3.0 mol $KMnO_4$?

$$(3.0\,mol\,KMnO_4)\left(\frac{5\,mol\,Cl_2}{2\,mol\,KMnO_4}\right)\left(\frac{6.022\times10^{23}\,molecules\,Cl_2}{mol\,Cl_2}\right) =$$

<u> 4.5×10^{24} molecules Cl_2 </u>

(c) To produce 35.0 g of $MnCl_2$, what mass of HCl will need to react?

$$(35.0\,g\,MnCl_2)\left(\frac{1\,mol\,MnCl_2}{125.8\,g\,MnCl_2}\right)\left(\frac{16\,mol\,HCl}{2\,mol\,MnCl_2}\right)\left(\frac{36.46\,g\,HCl}{mol\,HCl}\right) =$$

<u>   81.2 g HCl   </u>

(d) How many moles of water will be produced from 8.0 mol of $KMnO_4$?

$$(8.0\,mol\,KMnO_4)\left(\frac{8\,mol\,H_2O}{2\,mol\,KMnO_4}\right) =$$

<u>   32 mol H_2O </u>

(e) What is the maximum mass of Cl_2 that can be produced by reacting 70.0 g of $KMnO_4$ with 15.0 g of HCl?

$$(70.0\,g\,KMnO_4)\left(\frac{1\,mol}{158.0\,g}\right)\left(\frac{5\,mol\,Cl_2}{2\,mol\,KMnO_4}\right)\left(\frac{70.90\,g}{1\,mol\,Cl_2}\right) = 78.5\,g\,Cl_2$$

$$(15.0\,g\,HCl)\left(\frac{1\,mol}{36.46\,g}\right)\left(\frac{5\,mol\,Cl_2}{16\,mol\,HCl}\right)\left(\frac{70.90\,g}{1\,mol\,Cl_2}\right) = 9.12\,g\,Cl_2 \qquad$$ <u> 9.12 g Cl_2 </u>

Cl_2 **is clearly the limiting reactant.**

NAME **KEY**

SECTION _____ DATE _____

INSTRUCTOR _____

EXERCISE 12

Gas Laws

Show calculation setups and answers for all problems.

1. A sample of nitrogen gas, N_2, occupies 3.0 L at a pressure of 3.0 atm. What volume will it occupy when the pressure is changed to 0.50 atm and the temperature remains constant?

$(3.0 \text{ L } N_2)\left(\dfrac{3.0 \text{ atm}}{0.50 \text{ atm}}\right) =$

_____ 18 L _____

2. A sample of methane gas, CH_4, occupies 4.50 L at a temperature of 20.0°C. If the pressure is held constant, what will be the volume of the gas at 100.°C?

$(4.50 \text{ L } CH_4)\left(\dfrac{373 \text{ K}}{293 \text{ K}}\right) =$

_____ 5.73 L _____

3. The pressure of hydrogen gas in a constant-volume cylinder is 4.25 atm at 0°C. What will the pressure be if the temperature is raised to 80°C?

$(4.25 \text{ atm})\left(\dfrac{353 \text{ K}}{273 \text{ K}}\right) =$

_____ 5.50 atm _____

4. A 325 mL sample of air is at 720. torr and 30.°C What volume will this gas occupy at 800. torr and 75°C?

$(325 \text{ mL})\left(\dfrac{720. \text{ torr}}{800. \text{ torr}}\right)\left(\dfrac{348 \text{ K}}{303 \text{ K}}\right) =$

_____ 336 mL _____

5. A sample of gas occupies 500. mL at STP What volume will the gas occupy at 85.°C and 525 torr?

$(500. \text{ mL})\left(\dfrac{358 \text{ K}}{273 \text{ K}}\right)\left(\dfrac{760. \text{ torr}}{525. \text{ torr}}\right) =$

_____ 949 mL _____

6. A quantity of oxygen occupies a volume of 19.2 L at STP. How many moles of oxygen are present?

$$(19.2 \text{ L O}_2)\left(\frac{1 \text{ mol}}{22.4 \text{ L}}\right) =$$

_____0.857 mol_____

7. A 425 mL volume of hydrogen chloride gas, HCl, is collected at 25°C and 720. torr. What volume will it occupy at STP?

$$(425 \text{ mL HCl})\left(\frac{273 \text{ K}}{298 \text{ K}}\right)\left(\frac{720. \text{ torr}}{760. \text{ torr}}\right) =$$

_____369 mL_____

8. What volume would 10.5 g of nitrogen gas, N_2, occupy at 200. K and 2.02 atm?

$$PV = nRT \qquad V = \frac{nRT}{P}$$

$$V = (10.5 \text{ g N}_2)\left(\frac{1 \text{ mol}}{28.02 \text{ g}}\right)\left(\frac{0.0821 \text{ L–atm}}{\text{mol} \cdot \text{K}}\right)(200. \text{ K})\left(\frac{1}{2.02 \text{ atm}}\right) =$$

_____3.05 L_____

9. Calculate the density of sulfur dioxide, SO_2, at STP.

$$\left(\frac{64.06 \text{ g SO}_2}{1 \text{ mol}}\right)\left(\frac{1 \text{ mol}}{22.4 \text{ L}}\right) =$$

_____2.86 g/L_____

10. In a laboratory experiment, 133 mL of gas was collected over water at 24°C and 742 torr. Calculate the volume that the dry gas would occupy at STP.

$$(133 \text{ mL})\left(\frac{273 \text{ K}}{297 \text{ K}}\right)\left(\frac{(742 - 22) \text{ torr}}{760. \text{ torr}}\right) =$$

_____116 mL_____

11. A volume of 122 mL of argon, Ar, is collected at 50°C and 758 torr. What does this sample weigh?

$$(122 \text{ mL Ar})\left(\frac{273 \text{ K}}{323 \text{ K}}\right)\left(\frac{758 \text{ torr}}{760. \text{ torr}}\right)\left(\frac{1 \text{ L}}{1000 \text{ mL}}\right)\left(\frac{1 \text{ mol}}{22.4 \text{ L}}\right)\left(\frac{39.95 \text{ g}}{\text{mol}}\right) =$$

_____0.183 g_____

NAME **KEY**

SECTION ______ DATE ____________

EXERCISE 13

INSTRUCTOR ____________

Solution Concentrations

Show calculation setups and answers for all problems.

1. What will be the percent composition by mass of a solution made by dissolving 15.0 g of barium nitrate, $Ba(NO_3)_2$, in 45.0 g of water?

$$\left(\frac{15.0\text{ g }Ba(NO_3)_2}{60.0\text{ g solution}}\right)(100) =$$

$$\left(\frac{45.0\text{ g }H_2O}{60.0\text{ g solution}}\right)(100) =$$

$Ba(NO_3)_2$ **25.0%**

H_2O **75.0%**

2. How many moles of potassium hydroxide, KOH, are required to prepare 2.00 L of 0.250 M solution?

$$(2.00\text{ L})\left(\frac{0.250\text{ mol KOH}}{\text{L}}\right) =$$

0.500 mol KOH

3. What will be the molarity of a solution if 3.50 g of sodium hydroxide, NaOH, are dissolved in water to make 150. mL of solution?

$$\left(\frac{3.50\text{ g NaOH}}{150.\text{ mL}}\right)\left(\frac{1\text{ mol}}{40.00\text{ g NaOH}}\right)\left(\frac{1000\text{ mL}}{\text{L}}\right) =$$

0.583 M NaOH

4. How many milliliters of 0.400 M solution can be prepared by dissolving 5.00 g of NaBr in water?

$$(5.00\text{ g NaBr})\left(\frac{1\text{ mol}}{102.9\text{ g}}\right)\left(\frac{1\text{ mL}}{0.400\text{ mol}}\right)\left(\frac{1000\text{ mL}}{\text{L}}\right) =$$

121 mL

5. How many grams of potassium bromide, KBr, could be recovered by evaporating 650. mL of 15.0 percent KBr solution to dryness (d = 1.11 g/mL)?

$$(650.\text{ mL solution})\left(\frac{1.11\text{ g}}{\text{mL}}\right)\left(\frac{15.0\text{ g KBr}}{100.\text{ g solution}}\right) =$$

108 g KBr

6. How many milliliters of 12.0 M HCl is needed to prepare 300. mL of 0.250 M HCl solution?

$$(0.300\,\text{L})\left(\frac{0.250\,\text{mol}}{\text{L}}\right)\left(\frac{1\,\text{L}}{12.0\,\text{mol}}\right)\left(\frac{1000\,\text{mL}}{\text{L}}\right) =$$

<u>**6.25 mL 12.0 M HCl**</u>

7. A sample of potassium hydrogen oxalate, KHC_2O_4, weighing 0.717 g, was dissolved in water and titrated with 23.47 mL of an NaOH solution. Calculate the molarity of the NaOH solution.

$$KHC_2O_4 + NaOH \longrightarrow NaKC_2O_4 + H_2O$$

$$(0.717\,\text{g}\,KHC_2O_4)\left(\frac{1\,\text{mol}}{128.1\,\text{g}}\right)\left(\frac{1\,\text{mol NaOH}}{1\,\text{mol}\,KHC_2O_4}\right)\left(\frac{1}{0.02347\,\text{L NaOH}}\right) =$$

<u>**0.238 M NaOH**</u>

8. How many grams of hydrogen chloride are in 50. mL of concentrated (12 M) HCl solution?

$$(50.\,\text{mL})\left(\frac{1\,\text{L}}{1000\,\text{mL}}\right)\left(\frac{12\,\text{mol HCl}}{\text{L}}\right)\left(\frac{36.46\,\text{g}}{\text{mol}}\right) =$$

<u>**22 g HCl**</u>

9. A sulfuric acid solution has a density of 1.49 g/mL and contains 59 percent H_2SO_4 by mass. What is the molarity of this solution?

$$\left(\frac{1.49\,\text{g sol.}}{\text{mL}}\right)\left(\frac{59\,\text{g}\,H_2SO_4}{100.\,\text{g sol.}}\right)\left(\frac{1\,\text{mol}\,H_2SO_4}{98.09\,\text{g}\,H_2SO_4}\right)\left(\frac{1000\,\text{mL}}{\text{L}}\right) =$$

<u>**9.0 M H_2SO_4**</u>

10. Sulfuric acid reacts with sodium hydroxide according to this equation:

$$H_2SO_4 + 2\,NaOH \longrightarrow Na_2SO_4 + 2\,H_2O$$

A 10.00 mL sample of the H_2SO_4 solution required 18.71 mL of 0.248 M NaOH for neutralization. Calculate the molarity of the acid.

$$(0.01871\,\text{L NaOH})\left(\frac{0.248\,\text{mol NaOH}}{\text{L NaOH}}\right)\left(\frac{1\,\text{mol}\,H_2SO_4}{2\,\text{mol NaOH}}\right)\left(\frac{1}{0.01000\,\text{L}\,H_2SO_4}\right) =$$

<u>**0.232 M H_2SO_4**</u>

EXERCISE 14

Stoichiometry II

Show calculation setups and answers for all problems.

1. Use the equation to solve the following problems:

$$6\,KI + 8\,HNO_3 \longrightarrow 6\,KNO_3 + 2\,NO + 3\,I_2 + 4\,H_2O$$

(a) If 38 g of KI are reacted, how many grams of I_2 will be formed?

$$(38\,g\,KI)\left(\frac{1\,mol\,KI}{166.0\,g\,KI}\right)\left(\frac{3\,mol\,I_2}{6\,mol\,KI}\right)\left(\frac{253.8\,g\,I_2}{1\,mol\,I_2}\right) =$$

_____ 29 g I₂ _____

(b) What volume of NO gas, measured at STP, will be produced if 47.0 g of HNO_3 are reacted?

$$(47.0\,g\,HNO_3)\left(\frac{1\,mol\,HNO_3}{63.02\,g\,HNO_3}\right)\left(\frac{2\,mol\,NO}{8\,mol\,HNO_3}\right)\left(\frac{22.4\,L\,NO}{mol\,NO}\right) =$$

_____ **4.18 L NO** _____

(c) How many milliliters of 6.00 M HNO_3 will react with 1.00 mole of KI?

$$(1.00\,mol\,KI)\left(\frac{8\,mol\,HNO_3}{6\,mol\,KI}\right)\left(\frac{1\,L\,HNO_3}{6.00\,mol\,HNO_3}\right)\left(\frac{1000\,mL}{L}\right) =$$

_____ **222 mL 6.00 M HNO₃** _____

(d) When the reaction produces 8.0 mol of NO, how many molecules of I_2 will be produced?

$$(8.0\,mol\,NO)\left(\frac{3\,mol\,I_2}{2\,mol\,NO}\right)\left(\frac{6.022 \times 10^{23}\,molecules\,I_2}{mol\,I_2}\right) =$$

_____ **7.2 × 10²⁴ molecules I₂** _____

(e) How many grams of iodine can be obtained by reacting 35.0 mL of 0.250 M KI solution?

$$(0.0350\,L\,KI)\left(\frac{0.250\,mol\,KI}{L\,KI}\right)\left(\frac{3\,mol\,I_2}{6\,mol\,KI}\right)\left(\frac{253.8\,g\,I_2}{mol\,I_2}\right) =$$

_____ **1.11 g I₂** _____

2. Use the equation given to solve the following problems. All substances are in the gas phase.

$$N_2(g) + 3\,H_2(g) \longrightarrow 2\,NH_3(g)$$

(a) If 2.0 mol of H_2 react, how many moles of NH_3 will be formed?

$$(2.0\,\text{mol}\,H_2)\left(\frac{2\,\text{mol}\,NH_3}{3\,\text{mol}\,H_2}\right) =$$

$$\underline{\text{1.3 mol}\,NH_3}$$

(b) When 5.50 mol of N_2 react, what volume of NH_3, measured at STP, will be formed?

$$(5.50\,\text{mol}\,N_2)\left(\frac{2\,\text{mol}\,NH_3}{1\,\text{mol}\,N_2}\right)\left(\frac{22.4\,\text{L}\,NH_3}{\text{mol}\,NH_3}\right) =$$

$$\underline{\text{246 L}\,NH_3}$$

(c) What volume of NH_3 will be formed when 12.0 L of H_2, are reacted? All volumes are measured at STP

$$(12.0\,\text{L}\,H_2)\left(\frac{2\,\text{L}\,NH_3}{3\,\text{L}\,H_2}\right) =$$

$$\underline{\text{8.00 L}\,NH_3}$$

(d) How many molecules of NH_3 will be formed when 30.0 L of N_2 at STP react?

$$(30.0\,\text{mol}\,N_2)\left(\frac{1\,\text{mol}\,N_2}{22.4\,\text{L}\,N_2}\right)\left(\frac{2\,\text{L}\,NH_3}{1\,\text{mol}\,N_2}\right)\left(\frac{6.022\times10^{23}\,\text{molecules}\,NH_3}{\text{mol}\,NH_3}\right) =$$

$$\underline{1.61\times10^{24}\,\text{molecules}\,NH_3}$$

(e) What volume of NH_3, measured at 25°C and 710. torr, will be produced from 18.0 g of H_2?

$$(18.0\,\text{g}\,H_2)\left(\frac{1\,\text{mol}\,H_2}{2.016\,\text{g}\,H_2}\right)\left(\frac{2\,\text{mol}\,NH_3}{3\,\text{mol}\,H_2}\right)\left(\frac{22.4\,\text{L}}{\text{mol}}\right)\left(\frac{298\,\text{K}}{273\,\text{K}}\right)\left(\frac{760.\,\text{torr}}{710.\,\text{torr}}\right) =$$

$$\underline{\text{156 L}\,NH_3}$$

(f) If a mixture of 9.00 L of N_2 and 30.0 L of H_2 are reacted, what volume of NH_3 can be produced? Assume STP conditions.

3 L of H_2 are needed for each L of N_2
3 (9.00 L) = 27.0 L H_2 needed to react with 9.00 L N_2
Therefore, N_2 is the limiting reactant

$$(9.00\,\text{L}\,N_2)\left(\frac{2\,\text{L}\,NH_3}{1\,\text{L}\,N_2}\right) = \textbf{18.0 L}\,NH_3 \textbf{ can be produced.} \qquad \underline{\text{18.0 L}\,NH_3}$$

EXERCISE 15

Chemical Equilibrium

1. Consider the following system at equilibrium:

$$2\,CO_2(g) + 135.2\,kcal \rightleftharpoons 2\,CO(g) + O_2(g)$$

Complete the following table. Indicate changes in moles and concentrations by entering I, D, N, or? in the table (I = increase, D = decrease, N = no change, ? = insufficient information to determine).

Change or stress imposed on the system at equilibrium	Direction of shift, left or right, to reestablish equilibrium	Change in number of moles			Change in molar concentrations		
		CO_2	CO	O_2	CO_2	CO	O_2
a. Add CO	Left	I	I	D	I	I	D
b. Remove CO_2	Left	D	D	D	D	D	D
c. Decrease volume of reaction vessel	Left	I	D	D	I	I	I
d. Increase temperature	Right	D	I	I	D	I	I
e. Add catalyst	N	N	N	N	N	N	N
f. Add both CO_2 and O_2	?	I	?	I	I	?	I

2. Consider the reaction $PCl_5(g) \rightleftharpoons PCl_3(g) + Cl_2(g)$

At 250°C, PCl_5 is 45% decomposed.

(a) If 0.110 mol of $PCl_5(g)$ is introduced into a 1.00 L container at 250°C, what will be the equilibrium concentrations of PCl_5, PCl_3, and Cl_2?

(0.110)(0.45) = 0.050 mol PCl_5 decompose producing 0.050 mol each PCl_3 & Cl_2 leaving 0.110 − 0.050 = 0.060 mol PCl_5.

PCl_5 _____ **0.060 mol/L** _____

PCl_3 _____ **0.050 mol/L** _____

Cl_2 _____ **0.050 mol/L** _____

(b) What is the value of K_{eq} at 250°C?

$$K_{eq} = \frac{[PCl_3][Cl_2]}{[PCl_5]} = \frac{[0.050]^2}{[0.060]} = 4.2 \times 10^{-2}$$

K_{eq} _____ **4.2×10^{-2}** _____

3. For the reaction $H_2(g) + I_2(g) \rightleftharpoons 2\,HI(g)$, $K_{eq} = 0.17$ at 500 K. What concentration of $I_2(g)$ will be in equilibrium with $H_2 = 0.040$ M, $HI = 0.015$ M?

$$K_{eq} = \frac{[HI]^2}{[H_2][I_2]} = 0.17$$

$$[I_2] = \frac{[HI]^2}{(0.17)[H_2]} = \frac{(0.015)^2}{(0.17)(0.040)} = 0.030$$

_____ **0.033 M** _____

4. $CaCO_3$ has a solubility in water of 6.9×10^{-5} mol/L. Calculate the solubility product constant.

$$CaCO_3(s) \rightleftharpoons Ca^{2+}(aq) + CO_3^{2-}(aq)$$

$$K_{sp} = [Ca^{2+}][CO_3^{2-}] = (6.9 \times 10^{-5})(6.9 \times 10^{-5}) = 4.8 \times 10^{-9}$$

_____ **4.8 $\times$ 10^{-9}** _____

5. A 0.40 M HClO solution was found to have an H^+ concentration of 1.1×10^{-4} M. Calculate the value of the ionization constant. The ionization equation is $HClO \rightleftharpoons H^+ + ClO^-$.

$$K_a = \frac{[H^+][ClO^-]}{[HClO]} = \frac{(1.1 \times 10^{-4})^2}{(0.40)} = 3.0 \times 10^{-8}$$

_____ **3.0 $\times$ 10^{-8}** _____

6. Calculate (a) the H^+ ion concentration, (b) the pH, and (c) the percent ionization of a 0.40 M solution of $HC_2H_3O_2$ ($K_a = 1.8 \times 10^{-5}$).

(a) $K_a = \dfrac{[H^+][C_2H_3O_2^-]}{[HC_2H_3O_2]} = 1.8 \times 10^{-5}$

(a) _____ **2.7 $\times$ 10^{-3} M H$^+$** _____

(b) _____ **2.6** _____

(c) _____ **0.68%** _____

Let $x = [H^+] = [C_2H_3O_2^-]$

$$\frac{x^2}{0.40 - x} = 1.8 = 10^{-5}$$

Approximating, since x is very small

$x^2 = 1.8 \times 10^{-5}(0.40) = 7.2 \times 10^{-6}$

$x = 2.7 \times 10^{-3}$

(b) $pH = -\log[H^+] = -\log 2.7 \times 10^{-3} = 2.6$

(c) % ionization $= \left(\dfrac{[H^+]}{[HC_2H_3O_2]}\right)(100) = \left(\dfrac{2.7 \times 10^{-3}}{0.40}\right)(100) = 0.68\%$

EXERCISE 16

Oxidation-Reduction Equations I

Balance the following oxidation-reduction equations:

1. $3 \, P + \; 5 \; HNO_3 + \; 2 \, H_2O \longrightarrow \; 3 \, H_3PO_4 + \; 5 \, NO$

     $3 \, (P^0 \longrightarrow P^{5+} + 5e^-)$

     $5 \, (N^{5+} + 3e^- \longrightarrow N^{2+})$

2. $H_2SO_4 + \; 8 \, HI \longrightarrow H_2S + \; 4 \, I_2 + \; 4 \, H_2O$

     $S^{6+} + 8e^- \longrightarrow S^{2-}$

     $4 \, (2 \, I^- \longrightarrow I_2^0 + 2e^-)$

3. $KBrO_2 + \; 4 \, KI + \; 4 \, HBr \longrightarrow 5 \, KBr + \; 2 \, I_2 + \; 2 \, H_2O$

     $Br^{3+} + 4e^- \longrightarrow Br^-$

     $2 \, (2 \, I^- \longrightarrow I_2^0 + 2e^-)$

4. $6 \, Sb + \; 10 \, HNO_3 \longrightarrow 3 \, Sb_2O_5 + \; 10 \, NO + \; 5 \, H_2O$

     $3 \, (2 \, Sb^0 \longrightarrow \overset{10+}{Sb_2} + 10e^-)$

     $10 \, (N^{5+} + 3e^- \longrightarrow N^{2+})$

5. $3 \, NO_2 + \; 2 \, H_2O \longrightarrow \; HNO_3 + \; NO$

     $2 \, (N^{4+} \longrightarrow N^{5+} + 1e^-)$

     $N^{4+} + 2e^- \longrightarrow N^{2+}$

6. $3\,Br_2 +\ 8\,NH_3 \longrightarrow\ 6\,NH_4Br +\ \ N_2$

$3\,(Br_2^0 + 2e^- \longrightarrow 2\,Br^-)$

$2\,(N^{3-} \longrightarrow N_2^0 + 6e^-)$

7. $6\,KI +\ \ 8\,HNO_3 \longrightarrow\ 6\,KNO_3 +\ 2\,NO +\ \ 3\,I_2 +\ 4\,H_2O$

$3\,(2\,I^- \longrightarrow I_2 + 2e^-)$

$2\,(N^{5+} + 3e^- \longrightarrow N^{2+})$

8. $5\,H_2SO_3 +\ \ 2\,KMnO_4 \longrightarrow\ 2\,MnSO_4 +\ 2\,H_2SO_4 +\ \ K_2SO_4 +\ 3\,H_2O$

$5\,(S^{4+} \longrightarrow S^{6+} + 2e^-)$

$2\,(Mn^{7+} + 5e^- \longrightarrow Mn^{2+})$

9. $2\,K_2Cr_2O_7 +\ \ 2\,H_2O +\ \ 3\,S \longrightarrow\ 3\,SO_2 +\ \ 4\,KOH +\ \ 2\,Cr_2O_3$

$2\,(2\,Cr^{6+} + 6e^- \longrightarrow 2\,Cr^{3+})$

$3\,(S^0 \longrightarrow S^{4+} + 4e^-)$

10. $2\,KMnO_4 + 16\,HCl \longrightarrow\ \ 5\,Cl_2 +\ \ 2\,KCl +\ \ 2\,MnCl_2 +\ 8\,H_2O$

$5\,(2\,Cl^- \longrightarrow Cl_2^0 + 2e^-)$

$2\,(Mn^{7+} + 5e^- \longrightarrow Mn^{2+})$

EXERCISE 17

Oxidation-Reduction Equations II

Balance the following oxidation-reduction equations using the ion-electron method.

1. $2\,MnO_4^- + 10\,Cl^- + 16\,H^+ \longrightarrow 2\,Mn^{2+} + 5\,Cl_2 + 8\,H_2O$

    $2\,(Mn^{7+} + 5e^- \longrightarrow Mn^{2+})$

    $5\,(2\,Cl^- \longrightarrow Cl_2^0 + 2e^-)$

2. $3\,Ag_2S + 2\,NO_3^- + 8\,H^+ \longrightarrow 3\,S + 2\,NO + 6\,Ag^+ + 4\,H_2O$

    $3\,(S^{2-} \longrightarrow S_0 + 2e^-)$

    $2\,(N^{5+} + 3e^- \longrightarrow N^{2+})$

3. $ClO_4^- + 8\,I^- + 8\,H^+ \longrightarrow 4\,I_2 + Cl^- + 4\,H_2O$

    $(Cl^{7+} + 8e^- \longrightarrow Cl^-)$

    $4\,(2\,I^- \longrightarrow I_2^0 + 2e^-)$

4. $3\,Br_2 + 3\,H_2O \longrightarrow BrO_3^- + 5\,Br^- + 6\,H^+$

    $5\,(Br_2^0 + 2e^- \longrightarrow 2\,Br^-)$

    $(Br_2^0 \longrightarrow 2\,\overset{5+}{Br}O_3^- + 10e^-)$

5. $2\,MnO_4^- + 3\,HS^- + H_2O \longrightarrow 3\,S + 2\,MnO_2 + 5\,OH^-$

    $2\,(Mn^{7+} + 3e^- \longrightarrow Mn^{4+})$

    $3\,(S^{2-} \longrightarrow S^0 + 2e^-)$

6. $3\,H_2O_2 + IO_3^- \longrightarrow I^- + 3\,O_2 + 3\,H_2O$ (acid solution)

      $3\,(O_2 \longrightarrow O_2^0 + 2e^-)$

      $I^{5+} + 6e^- \longrightarrow I^-$

7. $Cl_2 + SO_2 + 2\,H_2O \longrightarrow SO_4^{2-} + 2\,Cl^- + 4\,H^+$ (acid solution)

      $Cl_2^0 + 2e^- \longrightarrow 2\,Cl^-$

      $S^{4+} \longrightarrow S^{6+} + 2e^-$

8. $5\,U^{4+} + 2\,MnO_4^- + 2\,H_2O \longrightarrow 2\,Mn^{2+} + 5\,UO_2^{2+} + 4\,H^+$ (acid solution)

      $5\,(U^{4+} \longrightarrow U^{6+} + 2e^-)$

      $2\,(Mn^{7+} + 5e^- \longrightarrow Mn^{2+})$

9. $6\,Fe(CN)_6^{3-} + Cr_2O_3 + 10\,OH^- \longrightarrow 6\,Fe(CN)_6^{4-} + 2\,CrO_4^{2-}$ (basic solution)

                                                  $+ 5\,H_2O$

      $6\,(Fe^{3+} + 1e^- \longrightarrow Fe^{2+})$

      $2\,Cr^{3+} \longrightarrow 2\,Cr^{6+} + 6e^-$

10. $2\,Cr(OH)_3 + 3\,O_2^{2-} \longrightarrow 2\,CrO_4^{2-} + 2\,H_2O + 2\,OH^-$ (basic solution)

      $2\,(Cr^{3+} \longrightarrow Cr^{6+} + 3e^-)$

      $3\,(O_2^{2-} + 2e^- \longrightarrow 2\,O^{2-})$

EXERCISE 18

Organic Chemistry I

1. Write structural formulas for the three different isomers of pentane, all having the molecular formula C_5H_{12}.

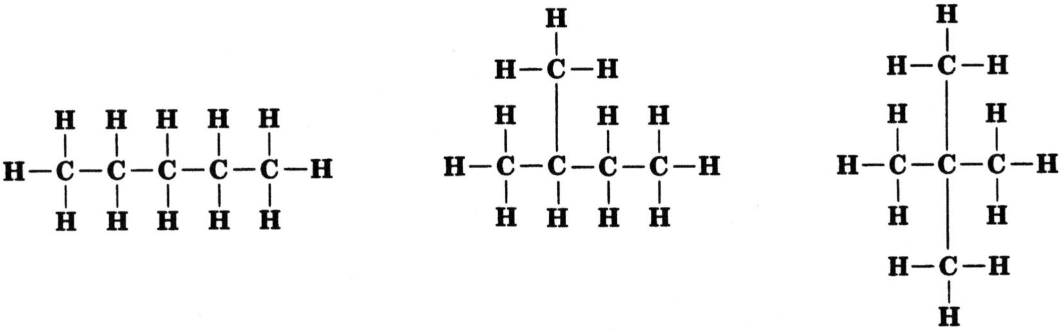

2. Write condensed structural formulas for (a) ethene, (b) propene, (c) three isomers of butene, (d) isomers of pentene, C_5H_{10}.

   (a) $CH_2=CH_2$ (b) $CH_3CH=CH_2$ (c) $CH_3CH_2CH=CH_2$ $CH_3CH=CH-CH_3$

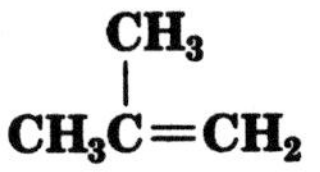

   $$CH_3\overset{\displaystyle CH_3}{\overset{|}{C}}=CH_2$$

   (d) $CH_2=CHCH_2CH_2CH_3$ 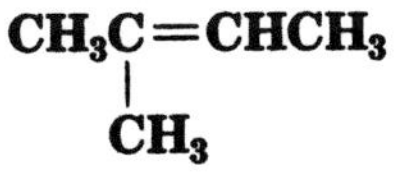 $CH_3\overset{}{C}=CHCH_3$
                                       $\quad\;\; |$
                                       $\quad\; CH_3$

   $CH_3CH=CHCH_2CH_3$

   $CH_2=\overset{}{C}CH_2CH_3$
   $\quad\;\; |$
   $\quad\; CH_3$

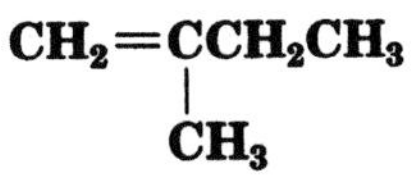

3. Write a balanced equation for the complete combustion of ethane.

   $2\,CH_3CH_3 + 7\,O_2 \longrightarrow 4\,CO_2 + 6\,H_2O$

4. Write condensed structural formulas for the five isomers of hexane, all having the molecular formula C_6H_{14}.

$$CH_3CH_2CH_2CH_2CH_2CH_3$$

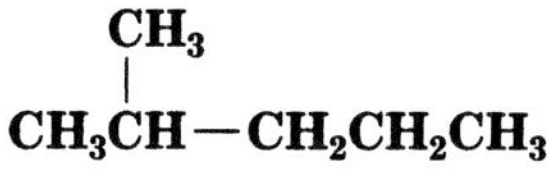

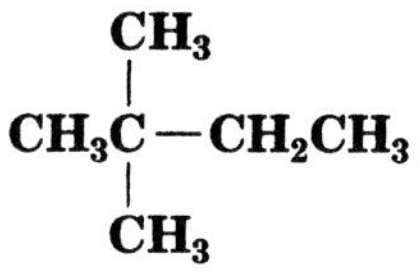

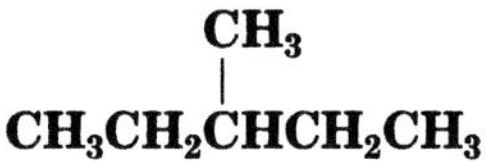

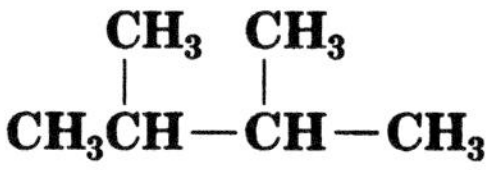

$$\overset{\displaystyle CH_3}{\underset{}{CH_3CH}}-CH_2CH_2CH_3$$

$$\overset{\displaystyle CH_3}{\underset{\displaystyle CH_3}{CH_3C}}-CH_2CH_3$$

$$\overset{\displaystyle CH_3}{\underset{}{CH_3CH_2CHCH_2CH_3}}$$

$$\overset{\displaystyle CH_3\ CH_3}{\underset{}{CH_3CH-CH-CH_3}}$$

5. Write condensed structural formulas for (a) acetylene, (b) 2-methylpropane (c) benzene, (d) *n*-octane.

(a) $H-C\equiv C-H$

(b) $\overset{\displaystyle CH_3}{\underset{}{CH_3CHCH_3}}$

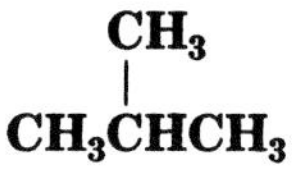

(c) 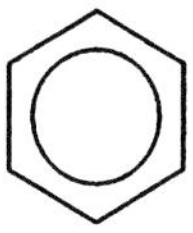 or

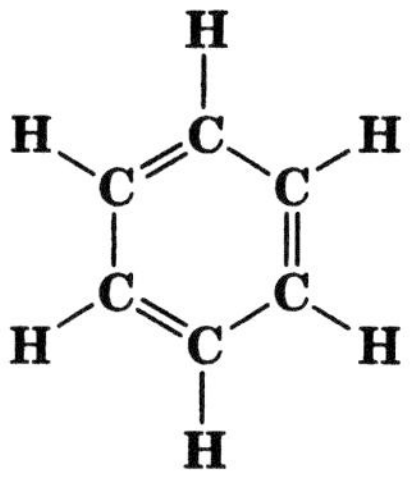

(d) $CH_3CH_2CH_2CH_2CH_2CH_2CH_2CH_3$

6. Write a balanced equation for the complete combustion of benzene.

$$2\ C_6H_6 + 15\ O_2 \longrightarrow 12\ CO_2 + 6\ H_2O$$

EXERCISE 19

Organic Chemistry II

1. Write condensed structural formulas for (a) ethyl alcohol, (b) acetone, (c) formaldehyde, (d) acetic acid.

   (a) CH_3CH_2OH

   (b) 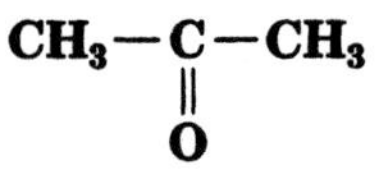 $CH_3-\overset{\displaystyle \underset{\|}{O}}{C}-CH_3$

   (c) $H-\overset{\displaystyle \underset{|}{H}}{C}=O$

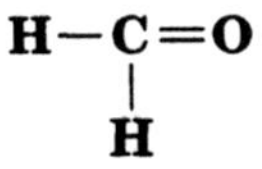

   (d) 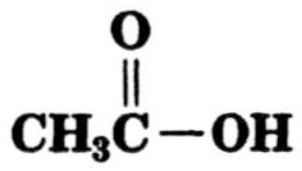 $CH_3\overset{\displaystyle \overset{O}{\|}}{C}-OH$

2. Write condensed structural formulas for the four different isomers of butyl alcohol, all having the formula C_4H_9OH.

   $CH_3CH_2CH_2CH_2OH$

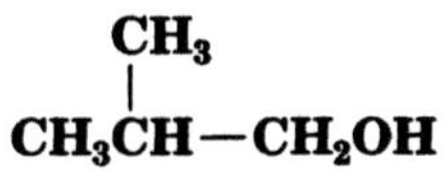

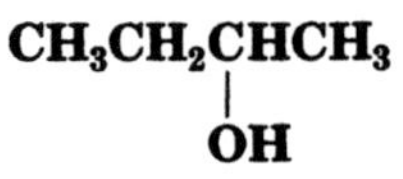

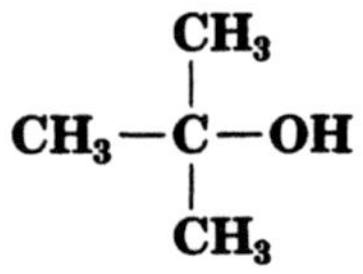

   $CH_3\overset{\displaystyle \overset{CH_3}{|}}{C}H-CH_2OH$

   $CH_3CH_2\overset{\displaystyle \underset{|}{OH}}{C}HCH_3$

   $CH_3-\overset{\displaystyle \overset{CH_3}{|}}{\underset{\underset{CH_3}{|}}{C}}-OH$

3. Write the name of the ester that can be derived from the following pairs of acids and alcohols:

Alcohol	Acid	Ester
Isopropyl alcohol	Acetic acid	**Isopropyl acetate**
Ethyl alcohol	Salicylic acid	**Ethyl salicylate**
Methyl alcohol	Stearic acid	**Methyl stearate**

4. The formula of butanoic acid is $CH_3CH_2CH_2COOH$. Write the structural formula for methyl butanoate.

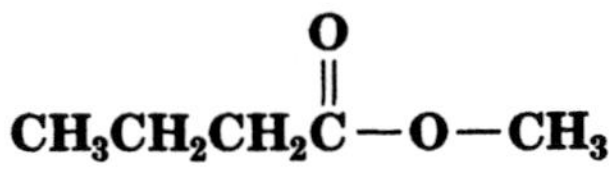

   $CH_3CH_2CH_2\overset{\displaystyle \overset{O}{\|}}{C}-O-CH_3$

5. The formula for benzoic acid is ⟨benzene ring⟩—COOH

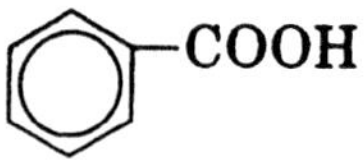

Write the structural formula for methyl benzoate.

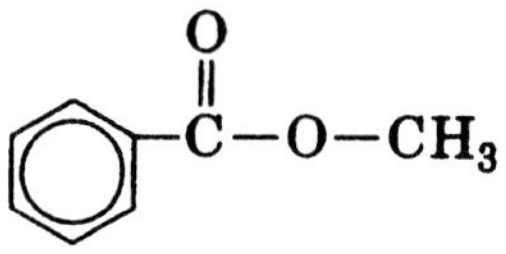

$$\text{⟨benzene ring⟩}-\overset{\displaystyle O}{\overset{\displaystyle \|}{C}}-O-CH_3$$

6. Identify the class of compound for each of the following:

(a) 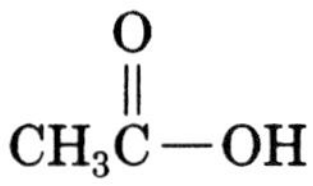
$$CH_3\overset{\displaystyle O}{\overset{\displaystyle \|}{C}}-OH$$

(b) CH_3CHCH_3
       |
       OH

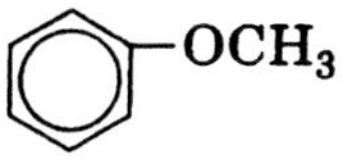

(a) _____Acid_____

(b) _____Alcohol_____

(c)
$$CH_3\overset{\displaystyle O}{\overset{\displaystyle \|}{C}}-OCH_3$$

(d) ⟨benzene ring⟩—COOH

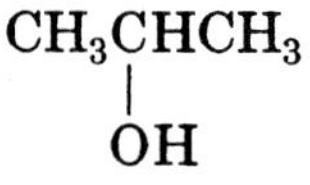

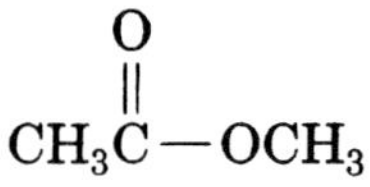

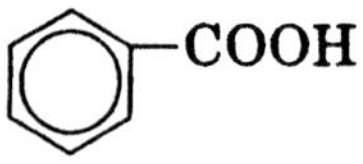

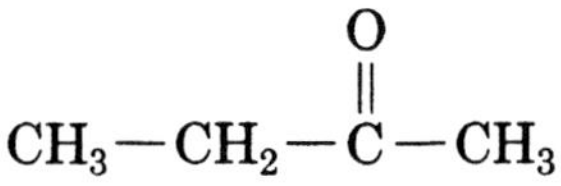

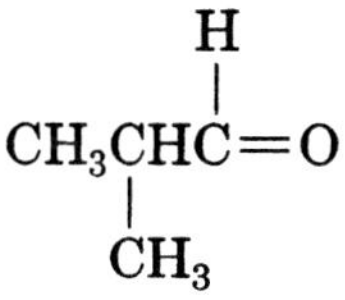

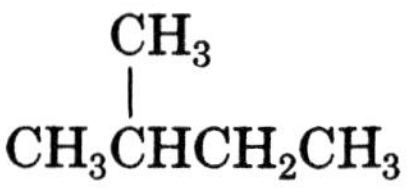

(c) _____Ester_____

(d) _____Acid_____

(e)
$$CH_3-CH_2-\overset{\displaystyle O}{\overset{\displaystyle \|}{C}}-CH_3$$

(f)
$$CH_3CH\overset{\displaystyle H}{\overset{\displaystyle |}{C}}=O$$
       |
       CH_3

(e) _____Ketone_____

(f) _____Aldehyde_____

(g)
$$\text{⟨benzene ring⟩}-\overset{\displaystyle O}{\overset{\displaystyle \|}{C}}-OCH_2CH_3$$

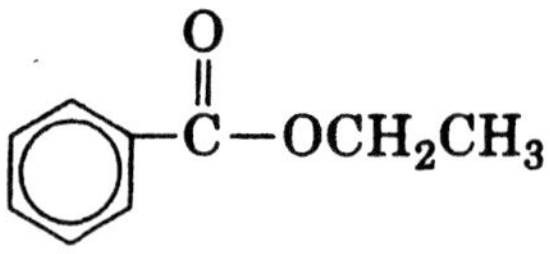

(h)
       CH_3
       |
$CH_3CHCH_2CH_3$

(g) _____Ester_____

(h) _____Hydrocarbon or Alkane_____

(i) CH_3COOH

(j) CH_3COOH_3

(i) _____Acid_____

(j) _____Ketone_____

(k) $CH_3CH(OH)CH_3$

(l) ⟨benzene ring⟩—OCH_3

(k) _____Alcohol_____

(l) _____Ester_____

Printed in the United States
37791LVS00001B/131

9 780471 468653